Robert J. Henry
Editor

Plant Conservation Genetics

Pre-publication
REVIEWS,
COMMENTARIES,
EVALUATIONS . . .

"**A**s efforts grow to conserve the world's plant diversity, understanding the genetic diversity of each of the target species will become increasingly important. Robert Henry has assembled outstanding authors at the cutting edge of conservation genetics. From *in situ* to *ex situ* situations, and from targeting populations to evaluating collections, this book reviews the current state of knowledge in the field and its potential importance to conservation efforts. Students of the field and all involved in the implementation of plant conservation efforts won't want to miss this invaluable discussion of a rapidly evolving field."

James S. Miller, PhD
William L. Brown Curator
And Director, William L. Brown
Center for Plant Genetic Resources,
Missouri Botanical Garden

"**T**his timely volume will be a great resource for research scientists, curators, and graduate and advanced undergraduate students in the plant genetic resource area. It covers both economically important food and fibre crops, and the issues of conservation of biodiversity, rare species, and the problems of fragmented habitats which are now to be found on every continent. Many of the chapters include information on the latest technical advances in plant conservation genetics, from collection to conservation methodologies and the use of the latest genomics techniques for sensible management of collections and habitats. The chapters by Nicole Rice and Glenn Bryan raise particularly important issues about the future utility of DNA banks and sequence information for conservation strategies in the freezer or potentially even 'in silico.' The final chapter by editor Robert Henry brings to light some of the challenges in using these techniques in an intelligent manner to enhance plant conservation strategies."

Ian Godwin, PhD, BAgrSc(Hons)
Associate Professor of Plant Molecular
Genetics, School of Land and Food
Resources, The University of Queensland

Plant Conservation Genetics

FOOD PRODUCTS PRESS®
Crop Science
Amarjit S. Basra, PhD
Editor in Chief

Plant-Derived Antimycotics: Current Trends and Future Prospects edited by Mahendra Rai and Donatella Mares

Concise Encyclopedia of Temperate Tree Fruit edited by Tara Auxt Baugher and Suman Singha

Landscape Agroecology by Paul A. Wojtkowski

Concise Encyclopedia of Plant Pathology by P. Vidhyasekaran

Molecular Genetics and Breeding of Forest Trees edited by Sandeep Kumar and Matthias Fladung

Testing of Genetically Modified Organisms in Foods edited by Farid E. Ahmed

Fungal Disease Resistance in Plants: Biochemistry, Molecular Biology, and Genetic Engineering edited by Zamir K. Punja

Plant Functional Genomics edited by Dario Leister

Immunology in Plant Health and Its Impact on Food Safety by P. Narayanasamy

Abiotic Stresses: Plant Resistance Through Breeding and Molecular Approaches edited by M. Ashraf and P. J. C. Harris

Teaching in the Sciences: Learner-Centered Approaches edited by Catherine McLoughlin and Acram Taji

Handbook of Industrial Crops edited by V. L. Chopra and K. V. Peter

Durum Wheat Breeding: Current Approaches and Future Strategies edited by Conxita Royo, Miloudi M. Nachit, Natale Di Fonzo, José Luis Araus, Wolfgang H. Pfeiffer, and Gustavo A. Slafer

Handbook of Statistics for Teaching and Research in Plant and Crop Science by Usha Rani Palaniswamy and Kodiveri Muniyappa Palaniswamy

Handbook of Microbial Fertilizers edited by M. K. Rai

Eating and Healing: Traditional Food As Medicine edited by Andrea Pieroni and Lisa Leimar Price

Handbook of Plant Virology edited by Jawaid A. Khan and Jeanne Dijkstra

Physiology of Crop Production by N. K. Fageria, V. C. Baligar, and R. B. Clark

Plant Conservation Genetics edited by Robert J. Henry

Introduction to Fruit Crops by Mark Rieger

Sourcebook for Intergenerational Therapeutic Horticulture: Bringing Elders and Children Together by Jean M. Larson and Mary Hockenberry Meyer

Agriculture Sustainability: Principles, Processes, and Prospects by Saroja Raman

Introduction to Agroecology: Principles and Practice by Paul A. Wojtkowski

Handbook of Molecular Technologies in Crop Disease Management by P. Vidhyasekaran

Handbook of Precision Agriculture: Principles and Applications edited by Ancha Srinivasan

Dictionary of Plant Tissue Culture by Alan C. Cassells and Peter B. Gahan

Handbook of Potato Production, Improvement, and Postharvest Management edited by Jai Gopal and S. M. Paul Khurana

Plant Conservation Genetics

Robert J. Henry
Editor

CRC Press
Taylor & Francis Group
Boca Raton London New York

CRC Press is an imprint of the
Taylor & Francis Group, an **informa** business

CRC Press
Taylor & Francis Group
6000 Broken Sound Parkway NW, Suite 300
Boca Raton, FL 33487-2742

© 20   by Taylor & Francis Group, LLC
CRC Press is an imprint of Taylor & Francis Group, an Informa business

Visit the Taylor & Francis Web site at
http://www.taylorandfrancis.com

and the CRC Press Web site at
http://www.crcpress.com

CONTENTS

ABOUT THE EDITOR

Robert James Henry, DSc, PhD, MSc, BSc, studies plants using molecular tools and is one of the leading plant scientists in Australia. His interests include Australian flora and plants of economic or social importance. Dr. Henry worked with the Commonwealth Scientific and Industrial Research Organisation (CSIRO) on fruit and vegetable biochemistry, the Queensland Department of Primary Industries (QDPI) as a cereal chemist, and the Queensland Wheat Research Institute as Senior Principal Scientist. His work has included assessment of quality in barley for use in malting and brewing and research on the impact of pre-harvest sprouting on the end-use quality of wheat; the study of DNA-based methods for identification of plants; the development of molecular markers for plant breeding; and the application of DNA technology to the improvement of the quality of crops and agricultural and food products.

Dr. Henry has written and edited several books on plant molecular biology and product quality and has published 200 refereed scientific papers and more than 300 national and international conference papers. He has been named one of the 100 most cited scientists in agriculture in international scientific literature over the past 20 years. He is a senior editor of the *Plant Biotechnology Journal* and a Fellow of the Royal Australian Chemical Institute (RACI). Dr. Henry received the Guthrie Award for contributions to cereal science in 2000, established the Australian Plant DNA Bank in 2001, and is a founder and director of Puragrain Pty Ltd.

doi:10.1300/5546_a

CONTRIBUTORS

Glenn Bryan, BSc, MSc, PhD, is a molecular geneticist at the Scottish Crop Research Institute, Invergowrie, Dundee, Scotland, United Kingdom.

Sally Dillon, BS, has a degree in agricultural science and is a research scientist in Australian Tropical Crops and Forages Collection, Agency for Food & Fibre Sciences, Queensland Department of Primary Industries, Brisbane, Queensland, Australia.

Mohammad Ehsan Dulloo, BSc (Hons), MSc, PhD, received an undergraduate degree with honors from Queen Mary College, University of London, UK, and graduate degrees in the conservation and utilization of plant genetic resources and conservation biology from the University of Birmingham, UK. Dr. Dulloo is currently working as a senior scientist for conservation at the International Plant Genetic Resources Institute, Rome, Italy.

Johannes M.M. Engels, DrIr, is the genetic resources management advisor at the International Plant Genetic Resources Institute (IPGRI), Rome Italy.

Robert Owen Makinson, BA, has degree in biology from Macquarie University and is the coordinator for the Centre for Plant Conservation, Botanic Gardens Trust, Sydney, NSW, Australia.

Tsukasa Nagamine, BS, PhD, has a degree in agriculture from Niigata University and a doctorate from Kyushu University and is Director of the Department of Crop Breeding, National Agriculture Research Center for Western Region, Fukuyama, Hiroshima, Japan.

Sir Ghillean T. Prance, BA, MA, DPhil, obtained his degrees from Oxford. He is a Fellow of the Royal Society (FRS) and has been awarded the Victoria Medal of Honour. He is Former Director of the Royal Botanic Garden, Kew.

doi:10.1300/5546_b

Nichole Rice, PhD, has a degree in science in agriculture Hons I and a PhD in agriculture from the University of Sydney. She is Curator at DNA Bank Pty Ltd, Lismore, NSW, Australia.

Maurizio Rossetto, BSc Hons, MSc, PhD, is Senior Research Scientist at the National Herbarium of New South Wales, Botanic Gardens Trust, Sydney, NSW, Australia.

Chuanqing Sun, PhD, has a degree in agronomy and is employed at the Department of Plant Genetics and Breeding, China Agricultural University, Beijing, China.

Imke Thormann, MSc, has a degree in natural sciences with a major in plant ecology and is Project Coordinator, Crop Wild Relatives Global Information System, Genetic Resources Science & Technology Group, IPGRI Headquarters, Vita dei Tre Danari, Italy.

Chapter 1

Plant Conservation Genetics: Importance, Options, and Opportunities

Robert J. Henry

INTRODUCTION

Plant conservation genetics is an emerging area of science. The narrowest definition of *plant conservation genetics* encompasses the analysis of wild populations of rare or endangered plants to understand the factors influencing the sustainability of the diversity within the species and of the species itself. Wider definitions include any study of plant genetics that could improve the understanding of the relationships between plants either wild or cultivated because this information is almost always likely to be useful in conservation of genetic diversity.

This book covers both in situ and ex situ conservation of all plant species whether they are of economic value or not. Conservation of plants in situ is the main option for plants with little value as ornamentals or in agriculture or forestry. Ex situ conservation is largely employed for economic species, especially those used in agriculture. However, the conservation of many economic species can be enhanced by conservation of wild relatives in situ.

Plant conservation genetics, along with most other areas of biology, has been greatly advanced by developments in molecular biology. The power of molecular genetic analysis and the increasing availability of cost-effective technologies allow plant conservation genetics to be applied much more widely in support of plant biodiversity conservation.

doi:10.1300/5546_01

IMPORTANCE OF PLANTS IN LIFE

Plants are biologically essential to life on earth. Plant conservation underpins the maintenance of animal life. Plants provide directly or indirectly basic necessities such as food and shelter for a wide range of organisms, including humans. Plant conservation genetics is key to the management of plant genetic resources for sustainable agriculture, forestry, and food production. Humans also value plant conservation for less obvious or immediate reasons. Plant diversity contributes both directly and indirectly to biodiversity in general. For many, the ability to experience the diversity of life forms is an important component of their quality of life. The contribution of plant conservation to the enrichment of life experiences is the primary motivation for the application of plant conservation genetics in nature conservation.

A few plant species, such as the cereals (Kettlewell and Henry, 1996), are the main components of human diets. A much larger number are minor sources of food. Many more species used in construction, clothing, and medicine are essential to many human communities.

ROLE OF GENETICS

Modern genetics builds on the understanding of inheritance as deduced by Mendel more than a century ago. It is empowered by knowledge of these mechanisms at the chemical level based on the double-helical structure of DNA elucidated over 50 years ago. Population genetics (Hartl, 1988) has become a well-developed discipline. Analysis of plant genomes at the DNA level allows investigation of genetic relationships between plants. Advances in techniques have made this approach much more practical in the past decade (Henry, 1997; Callow et al., 1997). The more recent development of genomics (the study of all the genes in organisms) promises to greatly enhance the future of plant conservation genetics.

The applications of genetic analysis in plants are manifold, including evolutionary biology, population genetics, and reproductive biology. Practical applications involve domestication and determination of gene function; determination of weediness in plant populations; and development of management strategies for conserving rare species, breeding plants, and identifying new uses for plants (Henry, 2001) and plant products.

OPTIONS AND STRATEGIES
FOR PLANT CONSERVATION

Plant genetic resources can be conserved in situ in wild populations or ex situ in collections such as seed banks. In situ conservation is essential for the survival of a very large proportion of plant species. The use of protected areas such as reserves and national parks needs to complement strategies for the native vegetation on other public and private lands.

Technical advances have enabled important gains in the ex situ conservation of plants. The ability to store seeds in seed banks is one. Maintaining living collections such as those in botanic gardens is another. A more recent advance is the DNA bank (Adams and Adams, 1992). These banks do not at present allow recovery of entire plants but do effectively conserve the genes themselves.

The choice of conservation method will depend on the species and populations under study. For example, populations conserved in situ may be vulnerable to changes in land use but should remain in situ for continued evolution of the species long term. The size of the population and the security of the site are key determinants.

OPPORTUNITIES FOR APPLICATION
OF PLANT CONSERVATION GENETICS

The applications of genetics in the conservation of plant biodiversity are summarized in Table 1.1. Genetic analyses of wild populations support their conservation by providing valuable input into management of in situ populations. This information may also assist in identifying appropriate material for ex situ conservation collections (Guarino et al., 1995). Analysis of such material also supports resource management by reducing duplication, ensuring protection of maximum diversity, and improving the plants. Genetic analyses provide the same benefits for domesticated plants as well.

To support plant diversity in wild plant populations, phylogenetic analysis based upon DNA sequencing defines evolutionary relationships between plants, and molecular analysis determines the genetic structures of plant populations. The extent of seed dispersal and pollen flow can be measured and the reproductive strategy of the plant established. These methods provide a basis for decisions to ensure conservation of a

TABLE 1.1. Application of Genetic Analysis in Plant Conservation

Applications of Genetic Analysis	Plant Population	Conservation Strategy
Conservation of wild plants	Wild plant population	In situ: Conserve diversity and evolution
Identification of targets for further collection		
Management of genetic resource collections	Plant genetic resource collections	Ex situ: Conserve diversity
Selection of germplasm for breeding		
Plant breeding	Domesticated plants	Ex situ: Maximize diversity in cultivated plants (especially in major food crops)

population that is viable in an evolutionary sense. Populations of sufficient size containing any genetically unique subpopulations can be protected when appropriate data are available.

Other methods support in situ and ex situ plant genetic resources as a broad base for crop production and forestry (Cooper et al., 2001; Young et al., 2000). These are focused at the genotype or gene level. Conservation of environmentally adapted genotypes provides healthy genetic backgrounds for plant breeding. Conservation of specific genes or alleles may allow enhancement of these genotypes in plant breeding for specific attributes.

Plant genome analysis addresses some essential biological questions. Comparative genome analysis yields new insights into evolutionary biology (Thomas et al., 2003). Phylogenetic analysis based on genomics allows investigations from the level of individual gene evolution to chromosomal rearrangements (Bowers et al., 2003). Rapid evolutionary processes may be important in understanding ecology (Yoshida et al., 2003) and may operate over very small geographical ranges in response to environmental gradients (Kalender et al., 2000). Plant conservation genetics is central to all of these issues and is crucial to extending plant biodiversity analysis beyond species richness assessments (Specht and Specht, 1993) to include variation within species.

OVERVIEW OF PLANT CONSERVATION GENETICS

This book outlines plant conservation genetics for both ex situ and in situ approaches. Topics include plant genetic resource collections, botanic gardens, herbaria, DNA banks, the impact of habitat fragmentation, conservation of rare species, and molecular and genomic techniques. The emphasis is on the practical aspects of the strategies and the technologies for conserving plant biodiversity. The chapters have been written by specialists with direct experience working in each of these fields. Thus, many key issues are illustrated with examples drawn from personal experience.

From the perspective of the practical plant conservationist, this volume describes the role of in situ and ex situ conservation in seed banks, botanic gardens, herbaria, and DNA banks and highlights developments in several underlying technologies. For example, plant phylogenetic (Chase et al., 1993) and population genetic (Maxted et al., 1997) analyses have become more effective because of the increasingly powerful tools being developed for molecular genetic analyses (Henry, 2001). Complex legal and ethical issues shape the policies and procedures of organizations and individuals involved in plant conservation. In this book the authors provide a practical guide, including links to appropriate literature and theoretical discussion, to help conservationsists navigate effectively in this complex environment.

REFERENCES

Adams, R.P. and Adams, J.E. (Eds.) (1992). *Conservation of Plant Genes: DNA Banking and in Vitro Biotechnology.* London: Academic Press.

Bowers, J.E., Chapman, B.A., Rong, J., and Paterson, A.H. (2003). Unravelling angiosperm genome evolution by phylogenetic analysis of chromosome duplication events. *Nature* 422, 433-438.

Callow, J.A., Ford-Lloyd, B.V., and Newbury, H.J. (Eds.) (1997). *Biotechnology and Plant Genetic Resources Conservation and Use.* Wallingford: CABI.

Chase, M.W., Soltis, D.E., Olmstead, R.G., Morgan, D., Les, D.H., Mishler, B.D., Duvall, M.R., Price, R.A., Hills, H.G., Qiu, Y.-L., Kron, K.A., et al. (1993). Phylogenetics of seed plants: An analysis of nucleotide sequences from the plastid gene rbcL. *Annals of the Missouri Botanic Gardens* 80, 528-580.

Cooper, H.D., Spillane, C., and Hodgins, T. (Eds.) (2001). *Broadening the Genetic Base of Crop Production.* Wallingford: CABI.

Guarino, L., Rao, V.R., and Reid, R. (Eds.) (1995). *Collecting Plant Genetic Diversity Technical Guidelines.* Rome: CABI International Plant Genetic Resources Institute.

Hartl, D.L. (1988). *A Primer of Population Genetics,* Second Edition. Sunderland, MA: Sinauer Associates.

Henry, R.J. (1997). *Practical Applications of Plant Molecular Biology.* London: Chapman & Hall.

Henry, R.J. (Ed.) (2001). *Plant Genotyping: The DNA Fingerprinting of Plants.* Wallingford: CABI.

Kalender, R., Tanskanen, J., Immonen, S., Nevo, E., and Schulman, A.H. (2003). Genome evolution of wild barley *(Hordeum spontaneum)* by BARE-1 retroposon dynamics in response to sharp microclimate divergence. *Proceedings of the National Academy of Sciences, United States of America* 97, 6603-6607.

Kettlewell, P.S. and Henry, R.J. (Eds.) (1996). *Cereal Grain Quality.* London: Chapman & Hall.

Maxted, N., Ford-Lloyd, B.V., and Hawkes, J.G. (Eds.) (1997). *Plant Conservation Genetics: The in Situ Approach.* London: Chapman & Hall.

Specht, A. and Specht, R.L. (1993). Species richness and canopy productivity of Australian plant communities. *Biodiversity and Conservation* 2, 152-167.

Thomas, J.W., Touchman, J.W., Blakesley, R.W., Bouffard, G.G., Beckstrom-Sternberg, S.M., Margulies, E.H., Blanchette, M., Siepel, A.C., Thomas, P.J., McDowell, J.C., et al. (2003). Comparative analysis of multi-species sequences from targeted genomic regions. *Nature* 424, 788-793.

Yoshida, T., Jones, L.E., Eliner, S.P., Fussmann, G.F., and Harison, N.G. (2003). Rapid evolution drives ecological dynamics in a predator-prey system. *Nature* 424, 303-306.

Young, A., Boshier, D., and Boyle, T. (Eds.) (2000). *Forest Conservation Genetics: Principles and Practice.* Wallingford: CABI.

Chapter 2

Techniques for ex Situ Plant Conservation

Imke Thormann
Mohammad Dulloo
Johannes Engels

INTRODUCTION

There are two major approaches to conservation of plant genetic resources: ex situ conservation and in situ conservation, including farming. The two approaches complement each other.

Ex situ conservation is defined in Article 2 of the Convention on Biological Diversity (CBD) (UNCED, 1992) as "the conservation of components of biological diversity outside their natural habitats." It involves the sampling, transference, and storage of target taxa from the collecting area and is generally used to safeguard species or populations that are at present or are potentially in danger of physical destruction, replacement, or genetic deterioration. The CBD definition of *in situ conservation* is the conservation of ecosystems and natural habitats and the maintenance and recovery of viable populations of species in the surroundings where the species have developed their distinctive properties (UNCED, 1992).

The techniques for ex situ conservation used today include seed banks, field gene banks, in vitro (slow-growth and crypreservation) storage, pollen banks, DNA storage, and botanic gardens.

This chapter explores ex situ conservation techniques, particularly field gene banks and in vitro and pollen storage. Seed storage, DNA storage, and botanic garden conservation are also described to illustrate their purposes, their advantages and disadvantages, and the types of species predominantly conserved. The establishment and

doi:10.1300/5546_02

management of collections and the complementarity of ex situ conservation in relation to in situ conservation are also briefly discussed.

The choice of which ex situ technique to use for germplasm storage is based on several factors. One is duration. The method of storage to conserve as much genetic diversity as possible for the future may be different from that for storage aimed at present and short-term use. Also the storage behavior of a species is decisive. The inherent longevity of seeds varies greatly, and species can show orthodox, recalcitrant, or intermediate storage behavior (Exhibit 2.1; Hong and Ellis, 1996). Other important factors are the human and financial resources and the institutional and technological capacities available.

HISTORICAL CONTEXT

Plant genetic resources have, sometimes unconsciously, been conserved since the beginning of agriculture, when plants and seeds were stored from one cycle of cultivation to the next and as people migrated from one area to another. As the human population grew and agriculture progressed, it became necessary to store seeds and plants for longer periods of time to expand cultivation and to colonize new land. More and more plants were domesticated, and the diversity of crops increased, largely due to human selection.

At the beginning of the past century new agricultural techniques allowed the development and cultivation of more productive but genetically less heterogenous crop varieties. In the 1920s and 1930s, N. I. Vavilov and later Jack Harlan and others began to notice that traditional crop varieties and highly diverse landraces were being lost

---

EXHIBIT 2.1. Seed Storage Behavior Categories

Orthodox: Seeds that can be dried to low moisture contents (3 percent for oily seeds and 7 percent for starchy seeds) without damage and be stored dry at low temperatures without losing their viability over long periods of time (Roberts, 1973).
Recalcitrant: Seeds that cannot withstand desiccation to moisture contents below 20 percent.
Intermediate: Seeds that can be dried to a moisture content of 10 to 12 percent, but further desiccation reduces viability and/or dry seeds are injured by low temperatures (Hong et al., 1996).

from cultivated fields around the world, as uniform modern varieties began to replace them (Scarascia-Mugnozza and Perrino, 2002). This led to the loss of genetic diversity, a process known as *genetic erosion*. The recognition of this loss prompted the collection of plant genetic resources worldwide and the establishment of the first gene banks. In 1967, the FAO/IBP Technical Conference on the Exploration, Utilization and Conservation of Plant Genetic Resources, organized by the International Biological Programme (IBP) and the Food and Agriculture Organization of the United Nations (FAO), recognized the need for long-term preservation, and ex situ long-term conservation was adopted as its most important method. However, the concern about the loss of genetic diversity continued to grow.

In 1974, the International Board for Plant Genetic Resources (IBPGR) was established, with the mission to coordinate an international plant genetic resources program in order to stem the loss of diversity. Collecting missions were accelerated, and gene banks were constructed and expanded at national, regional, and international levels. The number of ex situ storage facilities increased rapidly from 54 at the end of the 1970s to about 1,470 gene banks today, conserving more than 5.4 million accessions (Imperial College, 2002). These 5.4 million accessions are mostly plant genetic resources for food and agriculture. An additional 6 million accessions are conserved worldwide ex situ in ca. 2,000 botanic gardens in 148 countries (http://www.bgci.org.uk/botanic_gardens/index.html). The botanic garden collections consist mainly of wild, ornamental, and medicinal plants, but also forest species and crop species of essentially local significance. Species coverage in botanic gardens and gene bank collections is thus quite complementary.

Seed banking was the principal method used for conserving species ex situ. However, not all species can be conserved in seed banks. During the global collection and ex situ conservation efforts in the 1970s and 1980s, the first field gene banks were set up to conserve vegetatively propagated crops as well as those with long life cycles and those with recalcitrant seeds. In vitro storage techniques began to be developed as an alternative or complementary method for these types of species. In 1998, the FAO report on the state of the world's plant genetic resources for food and agriculture highlighted the need for further development of in vitro methods, cryopreservation methods, and complementary conservation strategies, employing both in situ and ex situ techniques. Since then considerable progress has been made in the development and use of cryopreservation techniques.

SEED AND ULTRA-DRY SEED STORAGE

Seed storage in gene banks is the most researched and usually the most efficient and effective form of long-term storage of plant germplasm of orthodox seeded species. Cereal crops and perhaps more than 80 percent of all other flowering plants produce orthodox seeds. Indeed, 90 percent of the accessions conserved in gene banks are maintained as seeds (FAO, 1998). Technical guidelines and standards are available, such as the gene bank standards published by FAO/ IPGRI (1994), guidelines for the design of seed storage facilities (Cromarty et al., 1982), and a protocol to determine seed storage behavior (Hong & Ellis, 1996). In general, this protocol involves the conservation of cleaned, healthy dried seeds (3 to 7 percent seed moisture content, depending on species) at subzero temperature but preferably at −18°C or cooler for long-term conservation (FAO/ IPGRI, 1994). For short- and medium-term conservation, seeds can be maintained at higher temperature (5°C to 10°C, depending on the objective of the conservation work). Seeds may be stored in cold rooms or in domestic deep chest freezers or refrigerators.

With regard to temperature, the effects on longevity from changes in seed storage temperature are very similar among diverse orthodox species. The relative benefit of a given reduction in temperature becomes less below a certain temperature.

Traditionally, seed banks have played their largest role in the conservation of domesticated plant varieties (Plucknett, 1987), although some agricultural seed banks such as those maintained by the U.S. National Plant Germplasm System have kept collections of non-domesticated species, particularly the wild relatives of crop plants. Now seed banks are gaining popularity also for the conservation of wild plant species (Schoen and Brown, 2001). The advantages and disadvantages to using seed banks as a method of conservation are summarized in Table 2.1.

Although the storage of orthodox seeds at −18°C is relatively simple, it can still be problematic for those gene banks for which lack of proper refrigeration equipment, unreliable electricity supply, poor maintenance practices, and high operating costs are major constraints.

TABLE 2.1. Advantages and Disadvantages of Seed Storage

Advantages	Disadvantages
Secure medium- to long-term conservation is feasible.	Building and equipping a gene bank is very expensive.
Seeds that are stored are generally free of pests and disease.	It is difficult to ensure adequate sampling for ex situ storage in gene banks.
Ex situ conservation of populations of most crops in gene banks ensures indefinite maintenance through cold storage and regeneration by exploitation of natural seed dormancy mechanisms.	Not all the variability within a species may be present, and gaps in intraspecific variability in gene banks may exist.
Seed banks require very little space to store many accessions of samples.	There is a risk of losing genetic diversity with each regeneration cycle.
Wide diversity can be conserved.	Genotypes may also be contaminated due to outcrossing and subsequent hybridization.
It is a convenient form for use and exchange of germplasm.	Regeneration is expensive and time consuming (especially for perennials).
No exposure occurs to environmental calamities such as diseases, drought, and floods.	Only species with orthodox seeds can be used.
	Evolutionary change is frozen because gene flow between wild, weedy, and cultivated plants is eliminated.

Many gene banks, especially those in developing countries, lack the funds and other resources needed to perform effectively. Research started in the late 1980s on so-called ultra-dry seed storage (Walters, 1998), which is based on the principle that desiccating seeds to much lower moisture contents than those generally used in standard procedures will allow them to be stored for longer periods at room temperature, thereby avoiding the requirement for refrigeration. Although the method seems to have fewer advantages than was initially expected, ultra-dry storage is still considered to be a practical, low-cost technique where no adequate refrigeration is available (Engelmann and Engels, 2002).

BOTANIC GARDEN CONSERVATION

Botanic gardens have traditionally focused on maintening species diversity (Heywood, 1991), particularly on wild species that are endangered in their natural habitat (Heywood, 1991, 1998; Wyse Jackson, 1998). Botanic gardens also specialize in medicinal and aromatic plants, ornamentals, and tropical species (Heywood, 1998; Rammeloo, 1998). Botanic gardens generally preserve their plant material as living collections consisting of only one or a few individuals per species. However, a survey of botanic gardens in 1994 found that 37 percent had seed storage facilities (Laliberté, 1997). Those species that cannot be stored as dried seed at low temperature are retained as tissue cultures (Engels and Engelmann, 2002).

DNA STORAGE

DNA banking is an emerging technique in genetic resources conservation. DNA extracts and DNA and RNA sequences are considered genetic resources, and they are now routinely extracted and conserved in DNA banks (Adams, 1997). The DNA bank of the Royal Botanic Garden Kew, for example, contains over 13,000 cryopreserved samples of plant genomic DNA (as of early 2002) (see http://www.rbgkew.org.uk/data/dnaBank/introduction.html). DNA extracts are mainly preserved for current and future research purposes, including phylogenetic, phylogeographic, and population studies; taxonomic and evolutionary research; and gene discovery and marker development. The advantage of storing DNA is that it is efficient and simple and overcomes many physical limitations and constraints that characterize other forms of storage. However, it is not yet possible to use DNA extracts to reconstruct any organism. Total genomic information is not the same as total genetic diversity, and limitations exist to the use of DNA banks for conservation, e.g., the fact that the genes will be separated from their phenotypes and DNA libraries cannot be screened by phenotype for useful agronomic traits (FAO, 1998).

FIELD GENE BANKS

Many plant species, especially of tropical origin, produce recalcitrant or intermediate seeds. Recalcitrant or intermediate seed-producing species are not as numerous as those that produce orthodox seeds, but many of them are economically important, e.g., oil palm *(Elaeis guineensis),* rubber *(Hevea brasiliensis),* durian *(Durio zibethinus),* coffee *(Coffea arabica),* cacao *(Theobroma cacao),* and coconut *(Cocos nucifera).*

In field gene banks, plant genetic resources are kept as live plants. According to FAO (1998), ca. 527,000 accessions are stored world-wide in field gene banks. Many species such as banana *(Musa* spp.), pineapple *(Ananas comosus),* sugarcane *(Saccharum* spp.), potato *(Solanum tuberosum),* and taro *(Colocasia esculenta)* reproduce mainly through vegetative means such as tubers, roots, suckers, and crowns. They rarely produce seeds and are highly heterozygous and therefore of limited utility for the conservation of particular genotypes. For these species and for species producing recalcitrant seeds, field gene banks have traditionally been the method of choice. Field gene banks can also be established as working collections of living plants for experimental and research purposes.

Field gene bank need a suitable climate and soil for the species. Preferably, they should be located within the boundaries of the centers of diversity of the particular crop and close to ongoing plant breeding (Hawkes et al., 2000).

The plants in a field gene bank undergo continuous growth and require the same constant care as in normal farming systems: adequate nutrition, pest and disease control, and irrigation. The "rule of thumb" is to use the same propagation techniques as the farmer, for example, not disrupting adapted clones through genetic segregation in a seed cycle. Local routine farming practices can be adapted for maintenance of a field gene bank.

The scale of management of field gene banks can be guaged from these examples. Oil palm genetic resources in Malaysia are planted at a density of 140 palms per hectare, and the collection from Nigeria alone occupies 200 ha (Engels and Wood, 1999). The manageable size for a sweet potato field bank, for example, was defined by Asian sweet potato workers to be a collection of 100 to 150 accessions (Rao and Schmiediche, 1996).

The International Plant Genetic Resources Institute (IPGRI) has published a training manual for the establishment and management of field gene banks (Mohd Said Saad and Ramanatha Rao, 2001). It is now producing new guidelines for the management of field and in vitro germplasm collections. The major advantages and disadvantages of field gene banks include the following.

The material in field gene banks is always readily available for characterization, evaluation, utilization, and research, whereas germplasm kept in the form of seeds or in in vitro cultures must first be germinated/regenerated and grown before it can be used. Plants with long life cycles often take many years to mature. Plants growing from seed can take more than ten years to mature and to start flowering, which is a problem if that germplasm is needed for immediate use. Maintaining such plants in a field gene bank is advantageous, as the plants will stay there for many years. Once they reach maturity (flowering and fruiting), they are in a ready-to-use form. They can be continuously evaluated, and crossing can be done at any time flowers are available.

The genotypes of many species that are cross-pollinated and vegetatively propagated are kept intact by growing the plants in the field in the form of clones. Breeders can then directly evaluate and select from the clones. Seeds from these plants usually give rise to segregating progenies. The segregation is even greater for polyploid plant species such as sweet potato *(Ipomoea batatas)* and potato *(Solanum tuberosum)*. Storing the seeds of these species might be relevant for gene conservation but not for genotype conservation. Furthermore, a much larger number of seeds must be stored in order to capture all possible genes from the particular genotypes. However, one will not be able to obtain the same genotypes by growing seeds.

As in farming, major constraints to field gene banks are pests and diseases, drought, flood, cyclones, vandalism, and theft. In addition, although field gene banks do not need costly equipment and sophisticated technology, it is normally more expensive to maintain plant genetic resources in the field than elsewhere. Living collections in field gene banks require large inputs of labor and vast areas of land to contain adequate samples of the genetic variability of the species. Therefore, shortages of land, labor, and financial resources can restrict the amount of genetic diversity conserved.

It is extremely difficult to keep vegetatively propagated plants free from viruses. Viral infections can lead to degeneration of clonal stocks and are a phytopathological threat (Hawkes et al., 2000). Reproduction from true seed, if this is possible, generally eliminates viral diseases, as only very few viruses are transmitted through pollen and true seed. As explained earlier, however, this method of reproduction is not appropriate for all genotypes. Finally, genetic erosion in some species or genetic groups may occur in field gene banks due to poor adaptation to the local environment (Table 2.2).

For these reasons, living collections in the field are not suitable for long-term conservation. Today, in vitro culture techniques, i.e., slow growth for medium-term and cryopreservation for long-term conservation, are developing as alternative and complementary methods for more efficient longer term conservation. The security and climatic independence of in vitro conservation can balance the risk and climatic specificity of field gene banks.

IN VITRO TECHNIQUES

Plant tissue culture and in vitro techniques are being refined particularly for vegetatively propagated crops, crop species with recalcitrant seeds, wild species that produce little or no seeds, and species with long life cycles. The basic aim is to introduce explants, i.e., small tissue pieces, from the donor plants into sterile culture and maintain them in a pathogen-free and controlled environment in a synthetic medium. The cultures can be stored under conditions of either slow or suspended growth. The slow-growth technique serves short- and medium-term conservation needs. For long-term storage of germplasm, in vitro cultured material is introduced to ultra-low temperatures so that virtually all cell activities are suspended. This is the cryopreservation technique.

The tissues utilized for culturing include meristems, shoot tips, axillary buds, and zygotic embryos. Although adventitious buds and

TABLE 2.2. Advantages and Disadvantages of Field Gene Banks

Advantages	Disadvantages
Suitable for species with recalcitrant and intermediate seeds	Vulnerable to changes in management practices
Convenient for characterization and evaluation	Susceptible to pests, diseases, and other natural and/or human-driven calamities such as drought, neglect, and war
Easily accessible for use	
Allow for conservation of particular genotypes	Limited amount of genetic diversity conserved
Possible to combine conservation and research/observation	High maintenance costs
	Not suitable for long-term conservation

somatic embryos derived from leaf, stem, root, or callus can also be used, such undifferentiated tissues should be conserved with caution because of their potential for somaclonal variation (Luan, 2001). Proper pathogen tests have to be done prior to storage as the in vitro cultured materials cannot be assumed to be automatically pathogen free. The advantages and disadvantages of in vitro cultivation are summarized in Table 2.3.

In vitro culture techniques such as meristem cultures, micro-propagation, and in vitro tuber induction have been developed in the past decades for over 1,000 species (George, 1996) and are continuing to be developed for more. For example, to facilitate conservation and reforestation, in vitro culture protocols were created for two threatened South African medicinal trees, *Ocotea bullata* and *Warburgia salutaris* (Kowalski and Van Staden, 2001). In vitro conservation protocols are now also available for a native tree of the Brazilian Atlantic Forest, *Cedrela fissilis* Vellozo (Da Nunes et al., 2003). Treatments for in vitro conservation for cutting buds of yam (*Dioscorea alata* L.) were recently tested by Borges et al. (2003).

In Vitro Slow Growth

The slow-growth method aims to minimize cell division and growth to increase longevity without genetic changes. The advantage is that the time between subculturing cycles is lengthened, thereby prolonging storage time and reducing maintenance costs. Strategies in which shoot-tip cultures and plantlets from meristems grow very slowly have the greatest application in genetic conservation. The stored

TABLE 2.3. Advantages and Disadvantages of in Vitro Conservation

Advantages	Disadvantages
Relatively easy alternative for medium- to long-term conservation of a large number of recalcitrant, sterile, or clonal species	Slow-growth method: risks of genetic instability; somaclonal variation and relatively high level of technology and high maintenance costs
Limited maintenance/monitoring required once material is placed in storage	Need to develop/adapt individual protocols for most species
Facilitated germplasm exchange: rapid in vitro multiplication of pathogen-free plant material	Difficulties with recalcitrant seed species
	Only a limited number of accessions can be stored

material is readily available for use and can easily be seen to be alive. Reduced vegetative growth of the stored material is induced through osmotic stress, limiting the availability of carbohydrate to suboptimal levels, keeping cultures at low temperature and/or in the dark, and/or incorporating growth retardants in the culture medium. These methods may be used singly or in combination. New techniques also under study include reduction of oxygen level using mineral oil layering or an environmental control; desiccation; and alginate encapsulation of the explants (Engelmann, 1999). The encapsulation of explants in alginate beads forms so-called synthetic seeds (synseed), which can be stored after partial dehydration. Apple and banana, for example, have been conserved successfully using synthetic seeds (Leela Sahijram and Rajasekharan, 1998).

The frequency of subculturing depends on the crop and its genetic variations. For example, the subculture frequency for clones of cassava (*Manihot* spp.) ranged from 12 to 18 months, depending on the cassava genotype (Roca et al., 2000). Banana shoot tips could be stored for up to 12 months under slow-growth conditions; several accessions have been maintained this way for ten years (Van den Houwe et al., 2000).

Experiments, carried out at ORSTOM (now IRD, Institut de recherche pour le développement) with slow-growth conditions characterized by a very low BA (6-benzyladenine) content of the culture medium and a subculturing interval of six months (Bertrand-Desbrunais et al., 1991), led to the successful storage of shoots of several genotypes of *Coffea congensis* Froehner, *C. canephora, C. liberica,* and *C. racemosa* Lour. for seven years without any loss (Dussert et al., 1997). In vitro shoot buds and callus cultures of date palm (*Phoenix dactylifera* L. cv Zaghlool) were successfully stored for 12 months (Bekheet et al., 2001).

Slow-growth techniques are routinely used for the conservation of vegetatively propagated species such as banana, plantain, cassava, potato, and sweet potato. Examples of very large in vitro collections are the cassava in vitro active collection at CIAT (Centro Internacional de Agricultura Tropical), Colombia, comprising more than 6,000 accessions (Roca et al., 2000), and the banana and plantain collection at the INIBAP (International Network for the Improvement of Banana and Plantain) Transit Center at the Katholieke Universiteit Leuven in Belgium. They conserve more than 1,000 *Musa* accessions as tissue cultures under slow-growth conditions. The Indian National Bureau for Plant Genetic Resources (NBPGR) conserves in vitro 1,300 accessions of vegetatively propagated plants, including fruits (strawberries, apples, kiwi, pear, and grapes) and medicinal and aromatic plants (R. Chaudhury, personal communication).

The main disadvantages of this method are the risk of somaclonal variation and the need to develop individual maintenance protocols for the majority of species. Once the culture is established, however, the accession is much safer than in a field bank, and the material can be exchanged more easily because it is less likely to carry pathogens than are cuttings. For example, slow-grown in vitro grape plants are kept in various gene banks as a back up to grape field collections (Toregrossa et al., 2000).

Cryopreservation

Cryopreservation of plant material offers long-term storage for germplasm that cannot be conserved as dry seed at subzero temperatures. Tissue cultures are conserved at very low temperatures, usually at −196°C in liquid nitrogen, to facilitate the arrest of mitotic and metabolic activities and thus to guarantee long-term preservation of germplasm in a genetically unaltered state. Cryopreservation protocols, like other in vitro culture techniques, need to be developed for each species. The basic requirements for cryopreservation protocols are that the plant material (1) survive the freezing procedure before storage and the thawing procedure after storage and (2) be regenerated into whole plants for use, conserving the genetic stability. Explants produced from apical and axillary meristems and somatic embryos have a high potential for genetic stability (Hawkes et al., 2000).

Most of the explants employed in cryopreservation (cell suspensions, calluses, shoot tips, embryos) contain high amounts of cellular water and are thus extremely sensitive to freezing injury. Cells have thus to be dehydrated artificially to protect them from damage caused by the crystallization of intracellular water into ice. The first two methods developed achieved this through freeze-induced dehydration, i.e., either ultra-rapid freezing or slow, stepwise freezing of cultures. The slow-freezing method depends for its success on protection by extracellular freezing. With the ultra-rapid freezing method, ice crystals form within the cells, but they are so small that they do not disrupt cell membranes. Often the material is treated with a cryoprotectant such as dimethylsulfoxide (DMSO), glycerol, ethylene glycol, proline, or sucrose, either singly or in combination. Cryoprotectants reduce the size and growth rate of ice crystals, lower the freezing point of intracellular contents, and enable cells to be subjected to very low temperatures without disruption of the cell membrane.

These classic techniques are generally operationally complex because they require sophisticated and expensive programmable freezers and a well-trained staff. They have been successfully applied to undifferentiated culture systems such as cell suspensions and calluses (Kartha and Engelmann, 1994; Withers and Engelmann, 1997) and to the apices of cold-tolerant species (Reed and Chang, 1997).

Significant progress has been made since the early 1990s with the development of efficient vitrification-based freezing protocols. Vitrification is the transition of water from the liquid phase directly into an amorphous glassy phase, avoiding the formation of ice crystals. In this procedure, cell dehydration is performed prior to freezing by exposure of samples to concentrated cryoprotective media and/or to air desiccation. Then follows rapid cooling. In all the new protocols, the dehydration step is critical to the survival of the explants, not the freezing step, as in the classic protocols (Engelmann, 2000). The vitrification-based procedures are more appropriate for complex organs, such as shoot tips and embryos, that contain a variety of cell types, each with unique requirements under conditions of freeze-induced dehydration. They are operationally less complex than the classic protocols and have a greater potential for broad applicability, as they require only minor modifications for different cell types (Engelmann, 1997). Engelmann (2000) has described the following seven procedures:

1. *Desiccation:* This technique is mainly used for zygotic embryos or embryonic axes extracted from seeds. The explants are dehydrated in the air current of a laminar airflow cabinet or in a flow of sterile compressed air or silica gel and then rapidly frozen by immersion into liquid nitrogen.

High survival rates were obtained for embryos of a number of recalcitrant and intermediate species, including oil palm (Ginibun, 1997), longan (Fu et al., 1990), arecanut (Raja et al., 2003), and *Dipterocarpus alatus* and *D. intricatus* (Krishnapillay et al., 1992). Ultra-rapid drying in a stream of compressed dry air allows freezing

of samples with a relatively high water content, thus reducing desiccation injury and improving survival in liquid nitrogen (Berjak
et al., 1989).

2. *Vitrification:* This technique involves the treatment of samples
first with cryoprotective substances. Then the tissue is desiccated
with a cocktail of vitrification solutions, followed by rapid freezing.
The vitrified material is thawed in 40°C water and the cryoprotectants
then removed. This procedure has been developed for apices, cell
suspensions, and somatic embryos of numerous species (Sakai, 1995,
1997). It has recently been applied successfully to embryogenic cultures of mango *(Mangifera indica)* (Huang et al., 2004) and *Arachis*
(Gagliardi et al., 2002), to carrot cell suspensions and protoplasts
(Yong and JunHui, 2003), and to dormant shoot tips of persimmon
(Ai and Luo, 2003). Matsumoto and Sakai (2003) have developed a
two-step vitrification protocol for the cryopreservation of axillary
shoot tips of grape that appears promising also for *Vitis* germplasm.

3. *Encapsulation-dehydration:* This method is successful for meristems and shoot tips of crops that are sensitive to direct desiccation.
Explants are encapsulated in alginate beads, pregrown in liquid medium enriched with sucrose for one to seven days, partially desiccated in the air current of a laminar airflow cabinet or with silica gel
down to a water content around 20 percent (fresh weight basis), and
then frozen rapidly. Survival rates are high, and growth recovery of
cryopreserved samples is generally rapid and direct, without callus
formation. The technique has been applied to apices of numerous
species of both temperate and tropical origins, e.g., *Citrus* species
(Santos and Stushnoff, 2002; Cho et al., 2002), *Dioscorea wallichii*
(Mandal et al., 1996), and *Saccharum officinarum* (Paulet et al.,
1993; Gonzalez-Arnao et al., 1996). It is applicable also to cell
suspensions and somatic embryos.

4. *Encapsulation-vitrification:* This procedure is a combination of
encapsulation-dehydration and vitrification. Samples are encapsulated in alginate beads and then subjected to freezing by vitrification.
It has been applied, e.g., to apices of lily (Matsumoto and Sakai,
1995; Hirai and Sakai, 2001), wasabi (Matsumoto et al., 1995),
Armoracia (Sakai, 1997), potato, strawberry, mint *(Mentha spicata),*
and Chinese yam *(Dioscorea* spp.) (Hirai and Sakai, 2001) and to
shoot tips of Troyer citrange *Poncirus trifoliata* (L.) Raf. × *Citrus
sinensis* (L.) Osbeck (QiaoChun et al., 2002).

5. *Pregrowth:* Samples are cultivated in the presence of cryo-protectants and then frozen rapidly by direct immersion in liquid nitrogen. The technique has been developed for *Musa* apices (Panis, 1995).

6. *Pregrowth-desiccation:* Explants are grown in the presence of cryoprotectants, dehydrated under the laminar airflow cabinet or with silica gel, and then frozen rapidly. This method has been applied to asparagus stem segments (Uragami et al., 1990), oil palm somatic embryos (Dumet et al., 1993), coconut zygotic embryos (Assy-Bah and Engelmann, 1992), and embryos of *Zizania texana* (Walters et al., 2002).

7. *Droplet-freezing:* Apices are pretreated with liquid cryopro-tective medium, placed on aluminium foil in minute droplets of fresh cryoprotectant, and frozen directly by rapid immersion in liquid ni-trogen. This technique has been applied to potato apices (Schaefer-Menuhr, 1996).

Operational simplicity is one important advantage of these new techniques, as they will be used mainly in tropical developing coun-tries where most genetic resources of problem species are located. Costs are low, and the samples take up very little space. After cryo-preservation, maintenance is reduced to mainly topping up the liquid nitrogen as there is no need for reculturing.

For many vegetatively propagated species, cryopreservation tech-niques are now sufficiently advanced to envisage their immediate uti-lization for large-scale experimentation in gene banks. Two examples are INIBAP and CIP (Centro Internacional de la Papa).

The research into improving the cryopreservation of banana and plantains is well advanced at INIBAP. As the objective is to cryo-preserve a rather large amount of accessions, the time necessary to prepare the material for freezing is a concern. Working with clumps of proliferating meristems—essentially little clusters of tiny shoots—a technician can prepare about 50 to 60 banana accessions per year to be frozen (IPGRI, 2003).

As of 2000, CIP had already cryopreserved 197 potato accessions. The aim is to store the whole potato base collection of ca. 4,000 ac-cessions in liquid nitrogen (Golmirzaie and Panta, 2000).

Methods for recalcitrant and intermediate seeded species are not as advanced because of the large number of mainly wild species (with very different characteristics) in this category and the comparatively limited research activities aimed at improving their conservation. Seeds and embryos of recalcitrant species also display various characteristics that make their cryopreservation difficult.

Because recalcitrant seeds do not have developmental arrests, it is essential to select the seeds at a precise stage. Selection can be difficult because variations in seed moisture content and maturity stage can occur between provenances, seed lots, and successive harvests. Also, seeds of many species are too large to be frozen directly, and embryos or embryonic axes have to be used. In some species the embryos or axes might be too large for cryopreservation or are not well defined (Engelmann, 1999). However, the most promising long-term conservation technology for recalcitrant seeds is the use of excised embryonic axes.

Considerable progress has been made in the cryopreservation of coffee. Different protocols are now available for coffee seeds, zygotic embryos, apices, and somatic embryos and have been applied to a relatively large range of species (Dussert et al., 2002). A major achievement is the first coffee cryobank now in place at CATIE (Centro Agronómico Tropical de Investigación y Enseñanza) in Costa Rica. Seeds of 79 accessions, representing the genetic diversity of the whole field gene bank and forming a core collection, are stored in liquid nitrogen (IPGRI, 2003). NBPGR, India, established a cryobank in 1996, which currently holds about 4,000 accessions of recalcitrant seeded species, including tea, jackfruit, litchi, citrus, oak, neem, almond, and cardamom (R. Chaudhury, personal communication).

Cryopreservation protocols are available for chayote, a recalcitrant tropical species (Abdelnour-Esquivel and Engelmann, 2002). Seed cryopreservation has been found to be an effective method of germplasm conservation for a range of *Piper* species (Decruse and Seeni, 2003). Research and development of protocols have also been started for tropical rare fruit species, as described by Normah et al. (2002) for some in Malaysia, many of which are recalcitrant.

The recent improvements in cryopreservation techniques make it now more and more possible to protect germplasm collections through long-term storage in liquid nitrogen and provide therefore a reliable back up for field and in vitro collections.

Cryopreservation research is not limited to vegetatively propagated, recalcitrant, or intermediate species. Orthodox seeded species have also been tested for their response to conservation in liquid nitrogen as an alternative to traditional cold storage. One example is the cryopreservation of 66 orthodox tropical Brazilian species (Salomao, 2002).

POLLEN STORAGE

Pollen grains are the mature male (haploid) gametophyte of higher plants. The longevity of the pollens of different species varies between minutes and years, depending on the species and on abiotic environmental conditions (Barnabas and Kovacs, 1997).

Pollen storage is common in breeding programs to bridge the gap between male and female flowering times and to improve fruit setting in orchards (Franklin, 1981; Towill, 1985; Alexander and Ganeshan, 1993). For example, the main use of coffee pollen is for breeding, because crosses sometimes have to be made between trees that do not flower simultaneously or that grow far apart (Walyaro and van der Vossen, 1977). Pollen is used for distributing germplasm to and exchanging it among locations because it rarely transfers pests and diseases (except for some viral diseases) and it is subjected to less stringent quarantine restrictions. Pollen storage also preserves the nuclear genes of germplasm. Finally, pollen is used in basic physiology, biochemistry, and fertility studies and in biotechnology studies involving gene expression, transformation, and in vitro fertilization (Towill and Walters, 2000).

Pollen is viable for only a relatively short period when conserved under classic storage conditions (partial desiccation followed by storage at subzero temperatures) and therefore is used only to a limited extent for germplasm conservation (Hoekstra, 1995). However, for some crops, the storage of pollen grains is possible in appropriate conditions and might be an interesting alternative for the long-term conservation of problematic species (IPGRI, 1996). When stored at the proper temperature and relative humidity (0°C to 10°C, 10 to 30 percent relative humidity, depending on species), the pollens of *Citrus* spp., *Cocos nucifera, Fragaria* sp., *Olea europea, Pinus silvestris, Pistachio atlantica, Pyrus malus,* and *Vitis vinifera* maintained their viability for more than one year (Barnabas and Kovacs, 1997). Walyaro and van der Vossen (1977) have obtained high pollen viability of *Coffea arabica* after more than two years by storing pollen under vacuum at –18°C. Yam pollen remained viable after being frozen at –80°C for two years (Ng and Daniel, 2000).

Pollen used for germplasm conservation should, however, remain viable for many years, and cryopreservation seems to be the most efficient method. For example, maize pollen could be dried to 50 percent of its original water content in an air current for one hour and then stored at –196°C in liquid nitrogen. Deep-frozen maize pollen can be used for fertilization after ten years of storage (Barnabas and Kovacs, 1997). Cryopreservation techniques have been developed for a large number of species, including mainly desiccation-tolerant pollen but also several desiccation-sensitive pollen (Towill, 1985; Bhat and Seetharam, 1993; Hanna and Towill, 1995), and cryobanks of pollen have been established for fruit tree species in several countries (Alexander and Ganeshan, 1993; Rajasekharan and Ganeshan, 1994; Rajasekharan et al., 1994, 1995; Ganeshan and Rajasekharan, 1995). At the Indian Institute for Horticultural Research (IIHR) in Bangalore, a pollen cryobank conserves about 300 accessions of fruit species, including citrus, mango, papaya, pomegranate, jack, and grape. *Pyrus* pollen is cryopreserved at the National Clonal Germplasm Repository in the United States (Reed et al., 2000). Ganeshan and Rajashekaran (2000) report the responses to cryopreservation experiments performed with pollen of 45 species belonging to 15 families, mainly horticultural species and fruit trees. With most of these species, protocols are optimized for establishing pollen cryobanks. The database software Polbase for digitizing accessions collected and maintained in a pollen cryobank was recently developed by Rajasekharan et al. (2003).

Knowledge of the cellular trait of pollen makes prediction of viability and storage characteristics is possible. The cell number in a pollen grain (bicellular or tricellular) at anthesis has been described for taxonomic families (Brewbaker, 1967). In many families all species possess the same pollen cell number. For example, all species of the Rosaceae contain bicellular pollen, and all species of the Compositae contain tricellular pollen. Some families contain genera with bicellular pollen and other genera with tricellular pollen. Most genera only contain a single type, but some rare exceptions occur. Generally, pollen that is bicellular at anthesis usually can be germinated in vitro, is shed at a lower moisture content, survives extensive desiccation, and is longer lived. Viability is more easily measured, and longevity can be greatly extended by desiccation and by lowering the storage temperature. Tricellular pollen usually has a high moisture content and survives only limited desiccation. The viability of tricellular pollen is difficult to measure, often being done by in situ growth in styles or seed set. Longevities of untreated or partially desiccated tricellular pollen at most noncryogenic temperatures are very short. However, there are exceptions, and various intermediate types exist.

Pollen storage also has several disadvantages. Many species do not produce amounts of pollen large enough for effective collection and processing. Transmission of organelle genomes via pollen is lacking, and sex-linked genes in dioecious species are lost. Pollen is an exhaustible reserve; provision must be made for periodic replenishment of the pollen banks. In this context it is obvious why pollen preservation is supplemental: the seed or clone must be conserved to yield the pollen. Multiple generations introduce the risk of population genetic problems such as loss of alleles through random drift or splitting of adaptive complexes. Only paternal material is conserved and regenerated, and, in order to utilize the germplasm, a recipient female plant is always needed for fertilization.

After consideration of the advantages and disadvantages of pollen conservation (summarized in Table 2.4), it can be concluded that storing pollen is useful in the framework of the establishment of complementary conservation strategies and may provide a means to overcome the difficulties involved in preserving recalcitrant-seeded species. This is particularly relevant because there is no correlation between seed storage behavior for a given species and the desiccation sensitivity of its pollen (Hoekstra, 1995).

TABLE 2.4. Advantages and Disadvantages of Pollen Storage

Advantages	Disadvantages
Viable alternative for species with recalcitrant seeds	Pollen of many species cannot be stored
Relatively easy procedure	Only paternal material conserved
Relatively low cost	Need to develop individual regeneration protocols to produce haploid plants
Facilitated exchange: relatively small quantity of material required for a single sample	
Pollen generally less likely to be infected by pathogens	Further research needed to produce diploid plants

GERMPLASM COLLECTION AND MANAGEMENT

Successful ex situ conservation depends not only on the proper choice and application of the techniques for storage, but also on the quality of the germplasm collected for conservation and on good management of the stored material. Guarino et al. (1995) provide comprehensive technical guidelines for collecting plant genetic diversity. Pence et al. (2002) provide specific guidelines on in vitro collecting techniques. IPGRI has published several technical guidelines and bulletins to aid gene bank curators in effective ex situ conservation. Among them are a decision guide for regenerating accessions in seed collections (Sackville Hamilton and Chorlton, 1997), a guide for the design and analysis of evaluation trials (IPGRI, 2000), a guideline for germplasm management at the accession level (Sackville Hamilton et al., 2002), a guidebook for genetic resources documentation (Painting et al., 1995), descriptor lists providing an international format and a universally understood "language" for plant genetic resources data (e.g., IPGRI, 2002), and guidelines for the safe movement of germplasm (e.g., Diekmann et al., 2003). The recent publication of a guide to effective management of germplasm collections (Engels and Visser, 2003) also touches on the political and economic aspects of gene bank conservation.

COMPLEMENTARY CONSERVATION

Each technique for ex situ conservation of plant material has its particular advantages and disadvantages, and together these techniques can and should complement each other (Dulloo et al., 1998; Damania, 1996; Maxted et al., 1997). If we add to these the various in situ conservation methods, such as conservation in protected areas, home gardens, and farms, there is quite a wide range of methods available for conservation. Although ex situ and in situ methods initially were characterized as having distinct agendas, they are in fact complementary. The complementarity of ex situ and in situ conservation has been stated in the CBD, and it is now well recognized that for any given gene pool a number of different approaches and methods are necessary for safe, efficient, and cost-effective conservation. No single method alone can conserve all the diversity of a gene pool.

For a root or tuber crop like cassava, germplasm is usually conserved in field gene banks. Cassava produces orthodox seeds, which could be stored dry for medium- and long-term conservation, although breeders prefer to use clonal material in their programs. Considering the immediate use value to breeders, field gene banks will probably continue to be an important element in cassava conservation, although in vitro conservation is already used as a back up to overcome the risk of losing the materials in the field collections. Cryopreservation techniques are being developed to ensure safe long- term conservation.

Similarly, INIBAP is developing a strategy for complementary conservation, including cryopreservation, in vitro methods, field gene banks, and on-farm conservation for banana and plantains.

In other situations, such as for wild relatives or forest tree species, in situ conservation in nature reserves and seed storage are more appropriate methods. Many indigenous wild plant populations approach even extinction. In such cases ex situ conservation and rare plant propagation become important complementary methods to help safeguard them.

Other elements that influence the conservation strategy include the biology and physiology of the plants, aspects of their management and use by humans, available infrastructure for conservation, number of accessions in a given collection and geographic sites, the purpose of conservation, and political and administrative policies (Withers, 1993; Dulloo et al., in press). As an example of a complementary strategy, Dulloo et al. (in press) discuss the options and factors influencing coconut germplasm and propose a conservation framework for coconuts. They conclude that such a strategy requires much effort and commitment from many stakeholders, who must work together with a common objective. Therefore, a proper enabling environment not only consisting of technical protocols and storage facilities but also including inter alia policy, finances, incentives, and, last but not least, good collaborative spirit, is crucial for the success of sustainable conservation of plant genetic resources.

REFERENCES

Abdelnour-Esquivel, A.; Engelmann, F. (2002). Cryopreservation of chayote (*Sechium edule* Jacq. SW.) zygotic embryos and shoot-tips from in vitro plantlets. *CryoLetters* 23, 299-308.

Adams, R.P. (1997). Conservation of DNA: DNA banking. In: Callow, J.A.; Ford-Lloyd, B.V.; Newbury, H.J. (eds.), *Biotechnology and Plant Genetic Resources:*

Conservation and Use. Biotechnology in Agriculture series: 19. CAB International, Wallingford, Oxon, UK, pp. 163-174.

Ai, P.F.; Luo, Z.R. (2003). Cryopreservation of dormant shoot-tips of persimmon by vitrification and plant regeneration. *Scientia Agricultura Sinica* 36(5), 553-556.

Alexander, M.P.; Ganeshan, S. (1993). Pollen storage. In: Chadha, K.L.; Adams, J.E. (eds.), *Advances in Horticulture,* Volume I—*Fruit Crops,* Part I. Malhotra Publishing House, New Delhi, India, pp. 481-496.

Assy-Bah, B.; Engelmann, F. (1992). Cryopreservation of mature embryos of coconut *(Cocos nucifera* L.) and subsequent regeneration of plantlets. *CryoLetters* 13, 117-126.

Bajaj, Y.P.S. (1995). Cryopreservation of plant cell, tissue, and organ culture for the conservation of germplasm and biodiversity. In: Bajaj, Y.P.S. (ed.), *Biotechnology in Agriculture and Forestry, Vol. 32: Cryopreservation of Plant Germplasm I.* Springer, Berlin, Germany, pp. 3-28.

Barnabas, B.; Kovacs, G. (1997). Storage of pollen. In: Shivanna, K.R.; Sawhney, V.K. (eds.), *Pollen Biotechnology for Crop Production and Improvement.* Cambridge University Press, Cambridge, UK, pp. 293-314.

Bekheet, S.A.; Taha, H.S.; Saker, M.M.; El-Bahr, M.K. (2001). In vitro long-term storage of date palm *(Phoenix dactylifera* L.). *Arab Universities Journal of Agricultural Sciences* 9, 793-802.

Berjak, P.; Farrant, J.M.; Mycock, D.J.; Pammenter, N.W. (1989). Homoiohydrous (recalcitrant) seeds: The enigma of their desiccation sensitivity and the state of water in axes of *Landolphia kirkii* Dyer. *Planta* 186, 249-261.

Bertrand-Desbrunais, A.; Noirot, M.; Charrier, A. (1991). Minimal growth in in vitro conservation of coffee *(Coffea* spp.): 1. Influence of low concentration of 6-benzyladenone. *Plant Cell, Tissue and Organ Culture* 27(3), 333-339.

Bhat, B.V.; Seetharam, A. (1993). Pollen storage and viability—An Indian perspective. In: Veeresh, G.K.; Umashankar, R.; Ganeshaiah, K.N. (eds.), *Pollination in the Tropics.* Proceedings of the International Symposium on Pollination in Tropics, August 8-13, 1993. IUSSI, Indian Chapter, GKVK, Bangalore, India, pp. 335-355.

Borges, M.; Meneses, S.; Vazquez, J.; Garcia, M.; Aguilera, N.; Infante, Z.; Rodriguez, A.; Fonseca, M. (2003). Conservacion in vitro de germoplasma de *Dioscorea alata* L. por crecimiento minimo. *Plant Genetic Resources Newsletter* 133, 8-12.

Brewbaker, J.L. (1967). The distribution and phylogenetic significance of binucleate and trinucleate pollen grains in the angiosperms. *American Journal of Botany* 54, 1069-1083.

Cho, E.G.; Hor, Y.L.; Kim, H.H.; Rao, Ramanatha V.; Engelmann, F. (2002). Cryopreservation of *Citrus madurensis* embryonic axes by encapsulation-dehydration. *CryoLetters* 23, 325-332.

Cromarty, A.S.; Ellis, R.H.; Roberts, E.H. (1982). *The Design of Seed Storage Facilities for Genetic Conservation*. Handbooks for Genebanks No. 1. International Board for Plant Genetic Resources, Rome, Italy.

Damania, A.B. (1996). Biodiversity conservation: A review of options complementary to standard ex situ methods. *Plant Genetic Resources Newsletter* 107, 1-18.

Da Nunes, E.; Benson, E.E.; Oltramari, A.C.; Araujo, P.S.; Moser, J.R.; Viana, A.M. (2003). In vitro conservation of *Cedrela fissilis* Vellozo (Meliaceae), a native tree of the Brazilian Atlantic forest. *Biodiversity and Conservation* 12, 837-848.

Decruse, S.W.; Seeni, S. (2003). Seed cryopreservation is a suitable storage procedure for a range of *Piper* species. *Seed Science and Technology* 31, 213-217.

Diekmann, M.; Sutherland, J.R.; Nowell, D.C.; Morales, F.J.; Allard, G. (eds.) (2003). *FAO/IPGRI Technical Guidelines for the Safe Movement of Germplasm No. 21. Pinus* spp. Food and Agriculture Organization, Rome, Italy, and International Plant Genetic Resources Institute, Rome, Italy.

Dulloo, M.E.; Guarino, L.; Engelmann, F.; Maxted, N.; Newbury, J.H.; Attere, F.; Ford-Lloyd, B.V. (1998). Complementary conservation strategies for the genus *Coffea*: A case study of Mascarene *Coffea* species. *Genetic Resources and Crop Evolution* 45, 565-579.

Dulloo, M.E.; Ramanatha Rao, V.; Engelmann, F.; Engels, J.M.M. (in press). Complementary conservation strategy for coconuts. In: Batugal, P.; Ramanatha Rao, V.; Oliver, J. (eds.), *Coconuts Genetic Resources*. International Plant Genetic Resources Institute-Regional Office for Asia, the Pacific, and Oceania (IPGRI-APO), Serdang, Selangor, Malaysia.

Dumet, D.; Engelmann, F.; Chabrillange, N.; Duval, Y. (1993). Cryopreservation of oil palm (*Elaeis guineensis* Jacq.) somatic embryos involving a desiccation step. *Plant Cell Reports* 12, 352-355.

Dussert, S.; Chabrillange, N.; Anthony, F; Engelmann, F.; Recalt, C.; Hamon, S. (1997). Variability on storage response within a coffee (*Coffea* spp.) core collection under slow growth conditions. *Plant Cell Reports* 16, 344-348.

Dussert, S.; Chabrillange, N.; Engelmann, F.; Anthony, F.; Vasquez, N.; Hamon, S. (2002). Crypresevation of *Coffea* (coffee). In: Towill, L.E.; Bajaj, Y.P.S (eds.), *Biotechnology in Agriculture and Forestry 50*. Cryopreservation of Plant Germplasm II. Springer, Germany, pp. 220-233.

Engelmann, F. (1997). In vitro conservation methods. In: Ford-Lloyd, B.V.; Newsbury, J.H.; Callow, J.A. (eds.), *Biotechnology and Plant Genetic Resources: Conservation and Use*. CABI, Wallingford, U.K., pp. 119-162.

Engelmann, F. (1999). Alternate methods for the storage of recalcitrant seeds—An update. In: Marzalina, Mo; Khoo, K.C.; Jayanthi, N.; Tsan, F.Y.; Krishnapillay, B. (eds.), *Proceedings of the IUFRO Seed Symposium 1998: Recalcitrant Seeds*. Forest Research Institute of Malaysia, Kepong, Malaysia, pp. 159-170.

Engelmann, F. (2000). Importance of crypreservation for the conservation of plant genetic resources. In: Engelmann, F.; Tagaki, H. (eds.), *Cryopreservation of Tropical Plant Germplasm. Current Research Progress and Application.* Japan International Research Center for Agricultural Sciences, Tsukuba, Japan/International Plant Genetic Resources Institute, Rome, Italy.

Engelmann, F.; Engels, J.M.M. (2002). Technologies and strategies for ex situ conservation. In: Engels, J.M.M.; Rao Ramanatha, V.; Brown, A.H.D.; Jackson, M.T. (eds.), *Managing Plant Genetic Diversity.* CABI Publishing, Wallingford, UK, pp. 89-103.

Engels, J.M.M.; Engelmann, F. (2002). Botanic gardens and agricultural genebanks: Building on complementary strengths for more effective global conservation of plant genetic resources. *Genetic Resources Newsletter* 131, 49-54.

Engels, J.M.M.; Visser, L. (eds.) (2003). *A Guide to Effective Management of Germplasm Collections.* IPGRI Handbooks for Genebanks No. 6. International Plant Genetic Resources Institute, Rome, Italy.

Engels, J.M.M.; Wood, D. (1999). Conservation of agrobiodiversity. In: Wood, D.; J.M. Lenné (eds.), *Agrobiodiversity.* CAB International, Wallingford, U.K., pp. 355-385.

FAO. (1998) *The State of the World's Plant Genetic Resources for Food and Agriculture.* Food and Agriculture Organization of the United Nations, Rome, Italy.

FAO/IPGRI. (1994). *Genebank Standards.* Food and Agriculture Organization/ International Plant Genetic Resources Institute, Rome, Italy.

Franklin, E.C. (ed.) (1981). *Pollen Management Handbook.* USDA Agriculture Handbook No. 587. USDA, Washington, DC.

Fu, J.R.; Zhang, B.Z.; Wang, X.P.; Qiao, Y.Z.; Huang, X.L. (1990). Physiological studies on desiccation, wet storage and cryopreservation of recalcitrant seeds of three fruit species and their excised embryonic axes. *Seed Science and Technology* 18, 743-754.

Gagliardi, R.F.; Pacheco, G.P.; Valls, J.F.M.; Mansur, E. (2002). Cryopreservation of cultivated and wild *Arachis* species embryonic axes using desiccation and vitrification methods. *CryoLetters* 23, 61-68.

Ganeshan, S.; Rajasekharan, P.E. (1995). Genetic conservation through pollen storage in ornamental plants. In: Chadha, K.L.; Bhattacharjee, S.K. (eds.), *Advances in Horticulture,* Volume 12—Ornamental Plants. Malhotra Publishing House, New Delhi, pp. 87-107.

Ganeshan, S.; Rajashekaran, R.K. (2000). Current status of pollen cryopreservation research—Relevance to tropical horticulture. In: Engelmann, F.; Takagi, H. (eds.), *Cryopreservation of Tropical Plant Germplasm—Current Research Progress and Applications.* Japan International Centre for Agricultural Sciences, Tsukuba/ International Plant Genetic Resources Institute, Rome, Italy, pp. 360-365.

George, E.F. (1996). *Plant Propagation by Tissue Culture. Part 2—In Practice,* Second Edition. Exegetics, Edington, UK.

Ginibun, F.C. (1997). *Effect of Speed of Desiccation and Prefreezing on Survival of Naked and Encapsulated Oil Palm* (Elaeis guineensis Jacq.) *Embryos in Liquid Nitrogen.* Graduation project, Fakulti Pertanian, Universiti Pertanian Malaysia.

Golmirzaie, A.M.; Panta, A. (2000). Advances in potato cryopreservation at CIP. In: Engelmann, F.; Takagi, H. (eds.), *Cryopreservation of Tropical Plant Germplasm—Current Research Progress and Applications*. Japan International Centre for Agricultural Sciences, Tsukuba/International Plant Genetic Resources Institute, Rome, Italy, pp. 250-254.

Gonzalez-Arnao, M.T.; Moreira, T.; Urra, C. (1996). Importance of pregrowth with sucrose and vitrification for the cryopreservation of sugarcane apices using encapsulation-dehydration. *CryoLetters* 17, 141-148.

Guarino, L.; Ramanatha Rao, V.; Reid, R. (eds.) (1995). *Collecting Plant Genetic Diversity: Technical Guidelines*. CAB International, Rome, Italy.

Hanna, W.W.; Towill, L.E. (1995). Long-term pollen storate. In: Janick, J.E. (ed.), *Plant Breeding Reviews,* Volume 13. John Wiley & Sons, New York, pp. 179-207.

Hawkes, J.G.; Maxted, N.; Ford-Lloyd, B.V. (2000). *The ex situ Conservation of Plant Genetic Resources*. Kluwer Academic Publishers, Dordrecht, the Netherlands.

Heywood, V.H. (1991). Developing a strategy for germplasm conservation in botanic gardens. In: *Tropical Botanic Gardens: Their Role in Conservation and Development*. Academic Press, New York, pp. 12-23.

Heywood, V.H. (1999). The role of botanic gardens in ex situ conservation of agrobiodiversity. In: Gass, T.; Frese, L.; Begemann, F.; Lipman, L. (eds.), *Implementation of the Global Plan of Action in Europe: Conservation and Sustainable Utilization of Plant Genetic Resources for Food and Agriculture*. Proceedings of the European Symposium on Plant Genetic Resources for Food and Agriculture. Braunschweig, Germany, June 30-July 4, 1998. International Plant Genetic Resources Institute, Rome, Italy, pp. 102-107.

Hirai, D.; Sakai, A. (2001). Recovery growth of plants cryopreserved by encapsulation-vitrification. *Bulletin of Hokkaido Prefectual Agricultural Experiment Stations* 80, 50-64.

Hoekstra, F. (1995). Collecting pollen for genetic resources conservation. In: Guarino, L; Rao, V.R.; Reid, R. (eds.), *Collecting Plant Genetic Diversity*. Technical guidelines. CAB International, Wallingford, UK, pp. 527-550.

Hong, T.D.; Ellis R.H. (1996). *A Protocol to Determine Seed Storage Behaviour*. IPGRI Technical Bulletin No. 1. International Plant Genetic Resources Institute, Rome, Italy.

Hong, T.D.; Linington, S.; Ellis, R.H. (1996). *Seed Storage Behaviour: A Compendium*. Handbooks for Genebanks No. 4. International Plant Genetic Resources Institute, Rome, Italy.

Huang, X.L.; Xiao, J.N.; Wu, Y.J.; Li, X.J.; Zhou, M.D.; Engelmann, F. (2004). Direct somatic embryogenesis induced from cotyledons of mango immature zygotic embryos. *In Vitro Cellular & Developmental Biology—Plant* 40(2), 196-199.

Imperial College. (2002). *Crop Diversity at Risk: The Case for Sustaining Crop Collections*. Dept. of Agricultural Science, Imperial College of Science, Technology and Medicine, Wye, U.K.

IPGRI. (1996). *Status Report on the Development and Application of in Vitro Techniques for the Conservation and Use of Plant Genetic Resources.* International Plant Genetic Resources Institute, Rome.

IPGRI. (2000). *Design and Analysis of Evaluation Trials of Genetic Resources Collections. A Guide for Genebank Managers.* IPGRI Technical Bulletin No. 4. International Plant Genetic Resources Institute, Rome, Italy.

IPGRI. (2002). *Descriptors for* Litchi (Litchi chinensis). International Plant Genetic Resources Institute, Rome, Italy.

IPGRI. (2003). *Thematic Report 2000-2001.* International Plant Genetic Resources Institute, Rome, Italy.

IPGRI/CIAT. (1994). *Establishment and Operation of a Pilot in Vitro Active Genebank.* IPGRI/CIAT, Rome and Cali.

Kartha, K.K.; Engelmann, F. (1994). Cryopreservation and germplasm storage. In: Vasil, I.K.; Thorpe, T.A. (eds.), *Plant Cell and Tissue Culture.* Kluwer, Dordrecht, the Netherlands, pp. 195-230.

Kowalski, B.; Van Staden, J. (2001). In vitro culture of two threatened South African medicinal trees—*Ocotea bullata* and *Warburgia salutaris. Plant Growth Regulation* 34, 223-228.

Krishnapillay, D.B.; Marzalina, M.; Pukttayacamee, P.; Kijkar, S. (1992). *Cryopreservation of* Dipterocarpus alatus *and* Diperocarpus intricatus *for Long Term Storage.* Poster presented at the 23rd International Seed Testing Association Congress. October 27-November 11, 1992. Buenos Aires, Argentina. Symposium Abstract No. 54:77.

Laliberté, B. (1997). Botanic garden seed banks/genebanks worldwide, their facilities, collections and network. *Botanic Gardens Conservation News* 2(9), 18-23.

Luan, H.Y. (2001). In vitro conservation and cryopreservation of plant genetic resources. In: Saad, M.S.; Ramanatha Rao, V. (eds.), *Establishment and Management of Field Genebank, a Training Manual.* IPGRI-APO, Serdang, pp. 54-58.

Mandal, B.B.; Chandel, K.P.S.; Dwivedi, S. (1996). Crypreservation of yam (*Dioscorea* spp.) shoot apices by encapsulation-dehydration. *CryoLetters* 17, 165-174.

Matsumoto, T.; Sakai, A. (1995). An approach to enhance dehydration tolerance of alginate-coated dried meristems cooled to −196°C. *CryoLetters* 16, 299-306.

Matsumoto, T.; Sakai, A. (2003). Cryopreservation of axillary shoot tips of in vitro-grown grape *(Vitis)* by a two-step vitrification protocol. *Euphytica* 131, 299-304.

Matsumoto, T.; Sakai, A.; Takahashi, T.; Yamada, K. (1995). Cryopreservation of in vitro grown meristems of wasabi *(Wasabia japonica)* by encapsulation-vitrification method. *CryoLetters* 16, 189-196.

Maxted, N.; Ford-Lloyd, B.V.; Hawkes, J.G. (1997). Complementary conservation strategies. In: Maxted, N.; Ford-Lloyd, B.V.; Hawkes, J.G. (eds.), *Plant Genetic Resources Conservation.* Chapman and Hall, London, pp. 15-39.

Mohd Said Saad; Ramanatha Rao, V. (eds.) (2001). *Establishment and Management of Field Genebank, a Training Manual.* IPGRI-APO, Serdang.

Ng, Q.N.; Daniel, I.O. (2000). Storage of pollens for long-term conservation of yam genetic resources. In: Engelmann, F.; Takagi, H. (eds.), *Cryopreservation of Tropical Plant Germplasm—Current Research Progress and Applications.* Japan International Centre for Agricultural Sciences, Tsukuba/International Plant Genetic Resources Institute, Rome, Italy, pp. 136-139.

Normah, N.M.; Clyde, M.M.; Cho, E.G.; Ramanatha Rao, V. (2002). Ex situ conservation of tropical rare fruit species. In: *Proceedings of the International Symposium on Tropical and Subtropical Fruits,* Volume 1. Cairns, Australia, November 26-December 1, 2000. *Acta Horticulturae* 575, pp. 221-230.

Painting, K.A.; Perry, M.C.; Denning, R.A.; Ayad, W.G. (1995). *Guidebook for Genetic Resources Documentation.* International Plant Genetic Resources Institute, Rome, Italy.

Panis, B. (1995). Cryopreservation of banana (*Musa* spp.) germplasm. Dissertationes de Agricultura, Katholieke Universiteit Leuven, Belgium.

Paulet, F.; Engelmann, F.; Glazmann, J.C. (1993). Cryopreservation of apices of in vitro plantlets of sugarcane (*Saccharum* sp.) hybrids using encapsulation-dehydration. *Plant Cell Reports* 12, 525-529.

Pence, V.C.; Sandoval, J.A.; Villalobos, V.M.; Engelmann, F. (eds.) (2002). In Vitro *Collecting Techniques for Germplasm Conservation.* IPGRI Technical Bulletin No. 7. International Plant Genetic Resources Institute, Rome, Italy.

Plucknett, D.L. (1987). *Gene Banks and the World's Food.* Princeton University Press, Princeton, NJ.

QiaoChun, W.; Batuman, O.; Li, P.; Bar-Joseph, M.; Gafny, R. (2002). A simple and efficient cryopreservation of in vitro grown shoot tips of Troyer citrange *Poncirus trifoliata* (L.) Raf. × *Citrus sinensis* (L.) Osbeck by encapsulation-vitrification. *Euphytica* 128, 135-142.

Raja, K.; Palanisamy, V.; Selvaraju, P. (2003). Desiccation and cryopreservation of recalcitrant arecanut (*Areca catechu* L.) embryos. *Plant Genetic Resources Newsletter* 133, 16-18.

Rajasekharan, P.E.; Ganeshan, S. (1994). Freeze preservation of rose pollen in liquid nitrogen: Feasibility, viability and fertility status after long-term storage. *Journal of Horticultural Science* 69, 565-569.

Rajasekharan, P.E.; Ganeshan, S.; Srinivasan, V.R. (2003). *Polbase: Digitalisation of Information Management System for Pollen Cryobanks.* National Seminar on Bioinformatics & Biodiversity Data Management. TBGRI, May 15-17, 2003, Thiruvananthapuram. Book of Abstracts, p. 10.

Rajasekharan, P.E.; Ganeshan, S.; Thamizharasu, V. (1995). Expression of trifoliate leaf character in *Citrus limonia* × *Poncirus trifoliolata* hybrids through cryostored pollen. *Journal of Horticultural Science* 70, 484-490.

Rajasekharan, P.E.; Rao, T.M.; Janakiram, T.; Ganeshan, S. (1994). Freeze preservation of gladiolus pollen. *Euphytica* 80, 105-109.

Rammeloo, J. (1998). *The Role of Botanic Gardens in Horticulture.* Paper presented at the XXVth International Horticultural Congress, August 2-7, 1998, Brussels, Belgium.

Rao, Ramanatha V.; Schmiediche, P. (1996). Conceptual basis for proposed approach to conserve sweet potato biodiversity. In: Rao, Ramanatha V. (ed.), *Proceedings of the Workshop on the Formation of a Network for the Conservation of Sweetpotato Biodiversity in Asia, Bogor, Indonesia, May 1-5, 1996.* International Plant Genetic Resources Institute, Rome.

Reed, B.M.; Chang, Y. (1997). Medium- and long-term storage of in vitro cultures of temperate fruit and nut crops. In: Razdan, M.K.; Cocking, E.C. (eds.), *Conservation of Plant Genetic Resources* in Vitro, Volume 1: General Aspects. Science Publishers Inc., Enfield, NH, pp. 67-105.

Reed, B.M.; DeNoma, J.: Chang, Y. (2000). Application of cryopreservation protocols at a clonal genebank. In: Engelmann, F.; Takagi, H. (eds.), *Cryopreservation of Tropical Plant Germplasm—Current Research Progress and Applications.* Japan International Centre for Agricultural Sciences, Tsukuba/International Plant Genetic Resources Institute, Rome, Italy, pp. 246-249.

Roberts, E.H. (1973). Predicting the storage life of seeds. *Seed Science and Technology* 1, 499-514.

Roca, W.M.; Debouck, D.; Escobar, R.H.; Mafla, G.; Fregene, M. (2000). Cryopreservation and cassava germplasm conservation at CIAT. In: Engelmann, F.; Tagaki, H. (eds.), *Cryopreservation of Tropical Plant Germplasm—Current Research Progress and Application.* Japan International Research Centre for Agricultural Sciences, Tsukuba/International Plant Genetic Resources Institute, Rome, Italy, pp. 273-279.

Sackville Hamilton, N.R.; Chorlton, K.H. (1997). *Regeneration of Accession in Seed Collections: A Decision Guide.* Handbooks for Genebanks No. 5. International Plant Genetic Resources Institute, Rome, Italy.

Sackville Hamilton, N.R.; Engels, J.M.M.; van Hintum, Th.J.L.; Koo, B; Smale, M. (2002). *Accession Management Trials of Genetic Resources Collections.* IPGRI Technical Bulletin No. 5. International Plant Genetic Resources Institute, Rome, Italy.

Sahijram, L.; Rajasekharan, P.E. (1998). Tissue culture strategies applicable to in vitro conservation of tropical fruit crops. In: *Tropical Fruits in Asia: Diversity, Maintenance, Conservation and Use.* Proceedings of the IPGRI-ICAR-UTFANET Regional Training Course on the Conservation and Use of Germplasm of Tropical Fruits in Asia, held at Indian Institute of Horticultural Research, May 18-31, 1997, Bangalore, India, pp. 113-119.

Sakai, A. (1995). Cryopreservation for germplasm collection in woody plants. In: Jain, S.; Gupta, P.; Newton, R. (eds.), *Somatic Embryogenesis in Woody Plants,* Volume 1. Kluwer, Dordrecht, the Netherlands.

Sakai, A. (1997). Potentially valuable cryogenic procedures for cryopreservation of cultured plant meristem. In: Razdan, M.K.; Cocking, E.C. (eds.), *Conservation of Plant Genetic Resources* in Vitro, Volume 1: General Aspects. Science Publishers Inc., Enfield, NH.

Salomao, A.N. (2002). Tropical seed species' responses to liquid nitrogen exposure. *Brazilian Journal of Plant Physiology* 14, 133-138.

Santos, I.R.I.; Stushnoff, C. (2002). Cryopreservation of embryonic axes of *Citrus* species by encapsulation-dehydration. *Plant Genetic Resources Newsletter* 131, 36-41.

Scarascia-Mugnozza, G.T.; Perrino, P. (2002). The history of ex situ conservation and use of plant genetic resources. In: Engels, J.M.M.; Rao Ramanatha, V.; Brown, A.H.D.; Jackson, M.T. (eds.), *Managing Plant Genetic Diversity.* CABI Publishing, Rome, Italy, pp. 1-22.

Schaefer-Menuhr, A. (1996). *Refinement of Cryopreservation Techniques for Potato. Final Report for the Period 1 September 1991-31 August 1996.* IBPGR Report. International Plant Genetic Resources Institute, Rome, Italy.

Schoen, D.J.; Brown, A.H.D. (2001). The conservation of wild plant species in seed banks. *Bioscience* 51(11), 960-966.

Torregrosa, L.; Bouquet, A.; Goussard, P.G. (2000). In vitro culture and propagation of grapevine. In: Roubelakis-Angelakis, K.A. (ed.), *Molecular Biology and Biotechnology of Grapevine.* Kluwer, Amsterdam, pp. 195-240.

Towill, L.E. (1985). Low temperature and freeze-/vacuum-drying preservation of pollen. In: Harthaa, K.K. (ed.), *Cryopreservation of Plant Cells and Organs.* CRC Press, Boca Raton, FL, pp. 171-198.

Towill, L.E.; Walters, C. (2000). Cryopreservation of pollen. In: Engelmann, F.; Takagi, H. (eds.), *Cryopreservation of Tropical Plant Germplasm—Current Research Progress and Applications.* Japan International Centre for Agricultural Sciences, Tsukuba/International Plant Genetic Resources Institute, Rome, Italy, pp. 115-129.

UNCED. (1992). *Convention on Biological Diversity.* United Nations Conference on Environment and Development, Geneva.

Uragami, A.; Sakai, A.; Magai, M. (1990). Cryopreservation of dried axillary buds from plantlets of *Asparagus officinalis* L. grown in vitro. *Plant Cell Reports* 9, 328-331.

Van den Houwe, I.; Panis, B.; Swennen, R. (2000). The in vitro germplasm collection at the Musa INIBAP Transit Centre and the importance of cryopreservation. In: Engelmann, F.; Tagaki, H. (eds.), *Cryopreservation of Tropical Plant Germplasm—Current Research Progress and Application.* Japan International-Research Centre for Agricultural Sciences, Tsukuba/International Plant Genetic Resources Institute, Rome, Italy, pp. 255-260.

Walters, C. (ed.) (1998). Ultra-dry seed storage. *Seed Science Research* 8, Suppl. 1.

Walters, C.; Touchell, D.H., Power, P.; Wesley-Smith, J.; Antolin, M.F. (2002). A cryopreservation protocol for embryos of the endangered species *Zizania texana. CryoLetters* 23, 291-298.

Walyaro, D.J.; van der Vossen, H.A.M. (1977). Pollen longevity and artificial cross-pollination in *Coffea arabica* L. *Euphytica* 26, 225-231.

Withers, L.A. (1993). New technologies for the conservation of plant genetic resources. In: Buxton, D.R.; Shibles, S.; Forsberg, R.A.; Blad, B.L.; Asay, K.H.; Paulsen, G.M.; Wilson, R.F. (eds.), *International Crop Science I. Proceedings of International Crop Science Congress,* Ames, Iowa, USA, July 14-22, 1992. Crop Science Society of America, Madison, WI, pp. 429-435.

Withers, L.A.; Engelmann, F. (1997). In vitro conservation of plant genetic resources. In: Altman, A. (ed.), *Biotechnology in Agriculture.* Marcel Dekker, Inc., New York, pp. 57-88.

Withers, L.A.; Engels, J.M.M. (1990). The test tube genebank—A safe alternative to field conservation. *IBPGR Newsletter for Asia and the Pacific* 3, 1-2.

Wyse Jackson, P. (1998). Botanic gardens: A revolution in progress. *World Conservation* (IUCN) 2/98, 14-15.

Yong, C.; JunHui, W. (2003). Cryopreservation of carrot (*Daucus carota* L.) cell suspensions and protoplasts by vitrification. *CryoLetters* 24(1), 57-64.

Chapter 3

Strategies Employed to Collect Plant Genetic Resources for ex Situ Conservation

Sally L. Dillon

INTRODUCTION

Two key strategies are employed to conserve plant species: in situ and ex situ conservation. As defined in Article 2 of the Convention on Biological Diversity, in situ conservation preserves ecosystems and natural habitats and maintains and recovers viable populations of species in the surroundings where they have developed their distinctive properties. In contrast, ex situ conservation of species preserves biological diversity outside of their natural habitats (UNCED, 1992). Put simply, in situ conservation maintains species in their natural habitats, while ex situ conservation removes species from their natural habitats. It must be emphasised that once a plant is removed from the natural habitat, it no longer experiences selective pressure from the natural environment (pest, disease, climatic), and thus the evolution of its potentially important genes ceases (Hoyt, 1988). In situ conservation therefore maintains the evolution of potentially important plant genes.

The seed type (if any) produced by plants determines whether in situ or ex situ strategies are employed to conserve its species. The seed produced by plants can be divided into two main types: orthodox and recalcitrant. Orthodox seeds undergo drying (moisture loss) during the seed maturation process and can maintain high viability even when dried down to very low moisture contents of 2 to 5 percent. More than 50 percent of the world's food species produce orthodox seeds, including all the major cereals and grasses (maize, wheat, rice,

doi:10.1300/5546_03

sorghum, forage species), many legumes and vegetables, cotton, and sunflower (Hoyt, 1988; Baskin and Baskin, 2001; Engelmann and Engels, 2002).

Recalcitrant seeds, on the other hand, do not undergo maturation drying and are shed with high moisture content. If seed moisture content is decreased below 20 percent, viability loss occurs, with eventual seed death. Species with recalcitrant seeds include mango, cocoa, avocado, rubber, tea, coffee, and breadfruit. They can be stored for only a matter of weeks or months and must be kept in a moist environment (Hoyt, 1988; Baskin and Baskin, 2001; Engelmann and Engels, 2002).

Conservation of species in situ is best suited to wild species and landrace germplasm and, in particular, to species that produce recalcitrant seed (see Prance, Chapter 7). In situ conservation preserves species in environments including protected areas (such as national and state parks, flora and fauna reserves), on farms of agricultural diversity, and in the home garden.

Ex situ conservation can preserve species as living plants (in botanic gardens, field gene banks, and tissue culture), as seed or tissue (long-term gene bank storage, tissue culture, and cryopreservation), and as DNA in plant DNA banks (Hoyt, 1988; Engelmann and Engels, 2002) (see also Thormann, et al., Chapter 2; Makinson, Chapter 5; and Rice, Chapter 6). Botanic gardens and field collections are particularly suited to those plant species that do not readily produce seeds or produce recalcitrant seeds, such as coconut and sugarcane (Hoyt, 1988). Tissue culture maintains species under slow-growth conditions, and this method is well suited to vegetatively propagated plant species and to those that produce recalcitrant seeds, including banana, potato, orchids, fruit trees, and some rare and endangered species (Sahijram and Rajasekharan, 1997; Engelmann and Engels, 2002). Cryopreservation involves snap freezing plant tissue in liquid nitrogen (−196°C) to bring all biological processes to a halt. When optimized for specific species, cryopreservation is a cost-effective, safe, long-term storage option best suited to recalcitrant and intermediate seed types such as potato, pear, and certain tree species and even to pollen from some horticultural species (Hoyt, 1988; Schäfer-Menuhr et al., 1997; Ganeshan and Rajashekaran, 2000; Panis et al., 2000).

DNA banks are a new form of genetic conservation in that they store the genomic DNA from plant species in a central location. Although DNA does not preserve a "living" system, it can assist in plant conservation by decreasing the demand of tissue and seed sampling from in situ plant populations and provide a central source of DNA for use in research programs (see Rice, Chapter 6). In situ conservation strategies are discussed in detail in Chapter 7, with ex situ conservation techniques discussed in Chapters 2, 5, and 6.

Plant genetic resources are collected for immediate and long-term uses, with both ensuring the conservation of important plant genes. Immediate uses of plant genetic resources include rehabilitation purposes, community projects, and scientific evaluations (Engels et al., 1995). Specific purposes for collecting plant species are to fill gaps in existing ex situ genetic resource collections so that the species and geographic diversity are adequately represented. Plants may also be collected for ecological studies of specific species, or there may be a need expressed for previously uncollected plant species. Opportunistic collecting occurs when species associated with the target populations are also collected.

Long-term ex situ conservation of plant species can prevent further genetic erosion and even extinction of species. Genetic erosion can result from habitat loss through urban spread, the construction of roads and dams, land clearing, and changes in agriculture. Detailed discussions of the causes of genetic erosion and the methods to measure the level of threat to species can be found in Engels et al. (1995), Brown and Marshall (1995), and Guarino (1995). Landrace species are being threatened in many countries around the world by the adoption of improved varieties. However, within Australian environments all crop varieties are of exotic origin (except Macadamia), so landraces are not a concern. The focus of this chapter is on the factors that drive the collection strategies used for ex situ conservation of orthodox seeds of Australian indigenous wild relatives of both crop and forage species for the Australian Tropical Crops and Forages Collection (ATCFC).

PLANNING THE COLLECTING MISSION

A successful collecting mission very much depends on the planning phase, as missions are expensive to undertake. The most effective use of time and resources before a mission is to plan it well. Factors to consider when planning include collection priorities, the optimum time for collection, appropriate sampling strategies, documentation, permissions required, team members, and the equipment needed. Above all others, the most important aspect before any plant-collecting mission is to set the objectives. Only after the objectives have been set can the priority target species and geographic range be determined, relevant collecting licenses and landholder permissions be obtained, and the number of collections needed to meet the objectives be determined. It is important also that the seed collecting not contribute to genetic erosion of or damage to the plant population or the natural ecosystem and that the seed be viable, representative, and adequately documented.

Setting Collection Priorities

The priorities that govern which species are collected include the threats facing a plant species and a plant species' level of genetic diversity and geographic distribution. Based on the level of threat (considering only Australian wild species), the highest collection priority would be endangered wild species (those at risk due to urbanization, agriculture, grazing, forestry, etc.) and the second would be wild species in general (in order to capture and centralize the variation available for scientific use in an ex situ collection) (Chapman, 1989).

Genetic relationships within a plant family do not necessarily reflect those based on morphology and taxonomy. Therefore, setting collection priorities based on taxonomy alone may not capture the representative genetic variation of the target species. To prioritize the importance of wild species in relation to their genetic relationships, Harlan and de Wet (1971) proposed a three-gene-pool system. Gene pool one (GP-1), the primary gene pool, contains the primary breeding material including crop species, progenitor, and wild and weedy forms of the species. There are no sterility barriers between these species, allowing straightforward gene transfer and fully fertile hybrids (Harlan and de Wet, 1971; von Bothmer and Seberg, 1995; Baskin and Baskin, 2001). Gene pool two (GP-2) is the secondary gene pool, and these species are able to cross with GP-1 species. Gene transfer between GP-1 and GP-2 species usually results in sterile hybrids; however, some fertility may be present (Harlan and de Wet, 1971; von

Bothmer and Seberg, 1995; Baskin and Baskin, 2001). The third (tertiary) gene pool (GP-3) contains species that are the extreme outer limit of potential genetic reach. They can be crossed with GP-1, but gene transfer is very difficult due to sterility barriers. Gene transfer can be obtained using radical measures; however, sterile hybrids are the result (Harlan and de Wet, 1971; Hoyt, 1988; Chapman, 1989).

Setting priorities based on the genetic (gene pool) system depends on the collection objectives. For breeding purposes, GP-1 species would have priority. For conserving and centralizing the genetic diversity across the plant family, species from GP-2 and GP-3 may be the target.

Advances in biotechnologies and crossing techniques have expanded the boundaries of the three-gene-pool classifications and made better use of GP-2 and GP-3 species for improving agronomic performance (Engels et al., 1995). Examples of such improvements include the following:

- Blue mould resistance in tobacco (*Nicotiana tabacum* L.) achieved through crossing with the Australian native species *N. goodspeedii* Wheeler, *N. debneyii* Domin, *N. excelsior* J. M. Black, *N. velutina* Wheeler, and *N. exigua* Wheeler (reviewed in Marshall and Broué, 1981).
- Improved yield in hybrids from cultivated rice (*Oryza sativa* L.) and the wild species *O. rufipogon* Griff. (Xiao et al., 1998).
- Resistance for the insect pest BPH (brown planthopper, *Nilaparvata lugens*) in cultivated rice obtained from the wild species *O. australiensis* Domin (Ishii et al., 1994).
- Pigeonpea *(Cajanus cajan)* lines with high seed protein from crosses with the Australian native *C. scarabaeoides* (L.) Thouars and many other Australian native *Cajanus* species identified as having pest and disease resistances (Reddy et al., 1997; Rao et al., 2003).

The true potential of wild relatives as sources of agriculturally important genes can only be assessed when the available collections represent their entire distributional and genetic ranges (Ladizinsky, 1989). Ultimately, there is no way of telling what tomorrow's needs may be or what plants might be able to fill them. Therefore, the more diversity that is conserved and made available for the future, the better the chances of fulfilling new demands (Engels et al., 1995).

Determining Specific Target Areas

The target areas for collecting are largely determined by the known distribution of species. The Australian Virtual Herbarium (www.chah. gov.au/avh/) contains historical records from all the Australian state and territory herbaria and is one of the best sources of information on the known distribution of plant species within Australia. Ex situ genetic resource collections, accessed though the Web database AusPGRIS (www.dpi.qld.gov.au/auspgris), are also good sources of information and contain data on the geographic range and seed availability of previously conserved species. Both of these databases can identify geographic regions that are rich in target species before decisions are made on where to focus collection activities.

Two criteria have been proposed to help prioritize collecting areas once the target species are determined. Where several species with similar characteristics occur in overlapping areas, the priority would be to collect in areas where multiple species occur. Where the target species are widespread over a large geographic range, then the priority may be to collect in areas where the greatest intraspecies genetic diversity may occur (Chapman, 1989).

Obtaining Permissions to Collect Seed

States and territories within Australia require plant collection licenses to be obtained prior to collecting seed from the wild. These licenses enable the states and territories to regulate the collection of plant germplasm and to protect plant populations and species from overcollection. Each license is specific to collection of the requested species from specified areas, so foreknowledge of the geographic range of the target species is required. Collecting licenses allow seed to be obtained from such areas as crown land, private properties, state parks, and some national parks. Additional licenses must be obtained for specific national parks in some states and territories and do require separate applications. Aboriginal lands also require special entry and separate collection licenses; the application forms are usually available from the state or territory land councils or the parks and wildlife service. It is important to apply for licenses well in advance of the proposed seed-collecting trip because it can take several months for an application to be processed after it is submitted.

Plant-collecting licenses contain specific restrictions and reporting requirements. One important requirement is that permission be obtained from landholders before entry onto private properties. A detailed report on collection activities is also required upon completion of any plant collecting missions, including specific geographic locations visited and the number of seeds or plants collected. Detailed information on how to obtain collecting licenses is available from the state and territory plant and wildlife departments listed in Exhibit 3.1.

Planning the Collecting Itinerary

After licenses and permissions have been obtained, a detailed collecting route/itinerary can be planned. From this plan, team members, vehicle needs, accommodation requirements, timing of the mission, and specific populations or environments of target species can be determined.

The team should be a small, multidisciplinary group, including a botanist if wild species will be collected or if the target species are

EXHIBIT 3.1. State/Territory Departments and Their Web Sites

Environment Australian Capital Territory	www.environment.act.gov.au
New South Wales National Parks and Wildlife Service	www.nationalparks.nsw.gov.au
Parks and Wildlife Commission of the Northern Territory	www.nt.gov.au/ipe/pwcnt/
Queensland Parks and Wildlife	www.epa.qld.gov.au/nature_conservation/plants
South Australian Department of Primary Industries and Resources	www.pir.sa.gov.au
Tasmanian Department of Primary Industries, Water and Environment	www.dpiwe.tas.gov.au
Victorian Department of Sustainability and Environment	www.dsc.vic.gov.au
Western Australian Department of Conservation and Land Management	www.calm.wa.gov.au/

unfamiliar. Valuable information about population localities and fruiting and seeding progress can be obtained from rangers and land-holders in the local areas, and the use of indigenous guides in remote or restricted areas is also a good idea. The number of members in the team and accommodation requirements will in part determine the type of vehicle needed. Helicopters have been used to collect from very remote or isolated areas, such as the Kimberley Ranges in Western Australia, that are not accessible by vehicle or on foot. Significantly more equipment (tents, cooking utensils, etc.) will be required for camping than for staying in a motel. The distance traveled during each day will be less if the team stays in a motel than if they camp out, as time must be set aside to return to the motel from the field. A good rule of thumb employed by the ATCFC team is a maximum distance of around 350 km per day if sampling across a large geographic region as this allows adequate time to collect seed at each site and return safely to overnight accommodation.

Unlike cultivated plants, wild species release their seed onto the ground once it is mature. The timely collection of mature seed is therefore important and can be determined from existing herbarium records. Narrowing the time period down to a few weeks in which to undertake the collecting mission can be achieved by talking to rangers and botanists working within the region, as they will have local knowledge of the growth stages of the target species. It is important to remember that variations in flowering time and seed set occur with changes in latitude of more than approximately 100 km. This variation in mature seed set was observed in the Northern Territory in 2002. Populations of native *Cajanus, Oryza,* and *Sorghum* growing around Tennant Creek (latitude $-19.44°S$) had shed most of their seed by the last week of April, while populations further north toward Adelaide River ($-13.28°S$) were in full seed set in the first week of May.

The annual or perennial nature of target species also affects the timing of the collecting trip. Perennial species of grasses and legumes such as *Cajanus, Glycine, Oryza,* and *Sorghum* set seed over a longer period than do annual species within the genus. This variation must be considered when determining the optimum time to collect. The timing and duration of the monsoonal rains throughout northern Australia can vary greatly from year to year, with short monsoon seasons resulting in seed set two to four weeks earlier in the year than longer or later monsoons. It is therefore important to establish the reproductive stage of development of the target species and the effects of both latitude and the annual/perennial nature of the species to ensure that the collecting trip coincides with adequate seed set on all target species.

If the target collecting region is close to your home base, there is the possibility of undertaking multiple short trips to coincide with mature seed set compared with a single extended mission that may be timed too early or late to collect mature seed samples. The collection mission of the ATCFC team in the Mount Isa region in 2000 targeted *Cajanus lanuginosis,* a species unrepresented in Australian ex situ genetic resource center collections. The trip was timed a few weeks early, and, rather than collect immature seed with reduced viability, the ATCFC returned to the identified *C. lanuginosis* sites at a later date and collected sufficient quantities of mature seed from a majority these sites.

Field Equipment

The basic field equipment used by the ATCFC is listed in Exhibit 3.2 and includes general items, field kits, and equipment to voucher specimens for lodgement with herbariums. Fine scale maps (no greater than 1:1,000,000) are important as they allow detailed travel routes to be determined to each collecting site and to the nearest town. Safety is of the utmost importance when travelling in remote areas, so share an itinerary of the daily activities and maintain regular phone contact with a base at set times.

Field Collection Strategies

A plant collector can sample only a fraction of the variation that occurs in nature. It is therefore crucial that this fraction be as large as possible and contain the maximum amount of useful variation to meet current and future needs (Namkoong, 1988; Brown and Marshall, 1995). Most wild species have developed highly localized patterns of gene distribution (localized diversity) in response to their natural environments, and these should also be sampled.

Genetic variation can be described by a large array of parameters, including allele and genotype frequencies, gene diversities, and heterozygosity levels (reviewed in Weir, 1990). Self-fertilizing and apomictic

EXHIBIT 3.2. Basic Equipment Required for Collecting Seed from the Field

Equipment	*Purpose*
General	
First aid kit	For unexpected emergencies
Detailed maps	To track where you are/where you are going
Wide brimmed hat & sunscreen	For sun protection
Insect repellent	To protect against insect bites
Long field clothes	To protect arms/legs from sun, scratches
Nonperishable food	At least two days' supply
Fresh water	Plenty of drinking water
Camping gear (if necessary)	Tent, sleeping gear, camp stove, etc.
Ventilated containers	For storing seed collections
Field kit	
Field data book	To record field data
Global Positioning System	To record longitude and latitude
Paper bags	To hold collected seed
Soil pH kit	To test soil pH
Camera	For locality and plant close-up shots
Pens, permanent markers	To write on tags, paper bags, in field book
Voucher specimens	
Secateurs	To minimize damage to plants when taking specimens
Newspaper	To place between each voucher specimen in presses
Day press	For storing daily voucher specimens
Tie-on tags	For labeling voucher specimens
Large field press	For storing dried voucher specimens
Rhizobium specimens	
Small hand trowel	For digging up roots and soil

Equipment	Purpose
Scalpel/sharp knife	For removing nodules
Magnifying hand lens	
Collecting vials with desiccant	For collecting samples

plant species closely resemble the parent, whereas outcrossing species differ from the parent and show within family diversity (Yonezawa and Ichihashi, 1989).

In general, outcrossing species are more diverse, although the frequency of alleles can vary significantly among populations (Loveless and Hamrick, 1984; Chapman, 1989). Alleles can therefore be categorized as those that are common versus rare and those that are widespread versus localized (Marshall and Brown, 1983). If sampling strategies focus on locally common alleles, then widespread common alleles will automatically be collected. Widespread rare alleles will be included in relation to the total number of individuals collected regardless of sampling strategy. The locally rare alleles are the most difficult to collect, and some would say that generally they are of little interest to other parties (Chapman, 1989). These rare alleles have adapted under localized selection pressures and may contain genes that are highly specialized against pests, disease, or environmental factors, and hence it could be argued that they are important to conserve ex situ.

To ensure that a collection is representative of the population, at least one copy of 95 percent of the alleles in the target population at a frequency of 5 percent (common alleles) or more should be collected (Marshall and Brown, 1975). A random collection of bulked seed from 30 individuals in fully outbreeding sexual species, 30 individuals in apomictic species, or 59 random individuals in a self-fertilizing species will ensure this. Marshall and Brown (1975) determined that 50 to 100 randomly selected individuals would collect all alleles with a frequency of 5 percent or more, and they set the benchmark at 50 random individuals for each population. A similar number of seeds should be collected from each individual (Marshall and Brown, 1975).

If less than 50 individuals form the population, then plants are lo-cally rare and seed collected from ten plants is adequate (Yonezawa, 1985). It is also important to remember that more alleles are likely to be conserved by collecting from a few plants from many sites than from many plants from a few sites. It may therefore be optimal to col-lect from populations that are widespread over geographic ranges and environments in order to capture the most genetic diversity.

If the target species or environment is uniform or sites occur over a broad geographic variation, then collect infrequently. If the target species and environment are variable, then collect more frequently. A combination of both is usually employed over the duration of a col-lecting mission. If there is low seed set from individuals within a pop-ulation, or you suspect low seed viability due to disease, predation, or seed immaturity, then the number of individuals collected should in-crease (Marshall and Brown, 1975). Ultimately, a sampling strategy needs to be defined that considers the distribution, life history, and genetic system of the target species so that the maximum amount of useful variation with a specified and limited number of samples can be obtained (Marshall and Brown, 1983).

The ATCFC aims to collect around 2,000 seeds from each target population. Combinations of the sampling strategies outlined previ-ously are used during the course of all collecting missions. Popula-tions found producing immature seed (less viable) are sampled only for voucher specimens with no seed harvested. This ensures that col-lections conserved in the ex situ genetic resource center are highly representative of the geographic and genetic ranges of the species.

DATA AND SEED COLLECTION

A collection is of little value if accompanied by insufficient field data. It is therefore crucial to record data for the end users of the germ-plasm. Specific locality and site data should include field taxonomic identification, detailed directions to the site, collector names, collec-tion number, photographs of both the population and the surrounding habitat (showing associated species), latitude, longitude, and altitude. Data on population dynamics, topography, soil properties, rhizobium sample, and any associated species should be detailed. Information that relates specifically to each population or collecting site (i.e., seed

predation, disease) should be noted. Steiner and Greene (1996) provide a detailed review of the collection data that need to be recorded
from the field, including habitat and accession descriptions. Table 3.1
summarizes the information that needs to be recorded at each collecting site. Preprinted carbon copy field books help ensure that all the
data are recorded at all collection localities. Figure 3.1 shows the field
book layout used by the ATCFC and describes a *Cajanus* accession
collected from the Northern Territory in 2002. Portable computers
(palm computers) are now being used to record collecting data in the
field. One advantage of these palm computers is that all collection
data can be downloaded straight into collection databases on computers off site.

Field collection numbers are assigned to each target population
from the field book, with all seed collected into paper bags labeled
with the field collection number and the date of collection. Collections are stored in a ventilated box/container within the vehicle until
completion of the trip. When collecting legume species, it is also important to collect a rhizobium/microsymbiont sample from the plant
nodules, roots, or soil at the site of collection. Sampling guidelines
for the collection of rhizobium/microsymbiont samples are detailed
in Date (1995).

Herbarium Voucher Specimens

Although a taxonomic identification is made in the field for each
collection, it is essential to verify the taxonomy by lodging a plant
cutting/voucher specimen with the state or territory herbarium where
you are collecting (and is often a condition of the collection license).
At least two voucher specimens should be taken from each population, showing both young and mature leaves, fruit/seeds, flowers, and
roots (where practical, and also in line with collecting license restrictions on root/soil collection to prevent potential disease spread).

To preserve voucher specimens correctly, each specimen should
have a tie-on tag labeled with the field book collection number. The
specimens should be pressed between newspapers in a field/day press
as soon as possible after collection to preserve plant structure. Species such as grasses are often very tall, and these can easily fit into the
day press by bending then into an M shape. The seeds of native plants

TABLE 3.1. Field Data That Must Be Recorded at Each Collecting Site

Data Field	Character	Example
Data collection	Collector	Collector names
		Field collection number
		Date of collection
	Collection site	Town/village/country
		Detailed road directions to site (include map reference)
		Latitude (using GPS in decimal format)
		Longitude (using GPS in decimal format)
		Altitude (using GPS in decimal format)
		Photographs of site, including associated species
Site physical description	Land use	Past/present such as cultivated, grazed, natural
	Soil properties	pH, texture (such as loam, sand), classification
	Slope	Incline (percent), aspect (such as S, SE, NW)
	Light	Open, ¼ shade, ½ shade, ¾ shade, shaded
	Natural moisture	Inundated all year, moist, seasonally dry, etc.
	Landform	Ridgetop, lower slope, rock outcrop, dry creek, etc.
Site vegetation description	Structure	Closed, open, sparse
	Formation group and class	Evergreen scrub, seasonal tall grass, etc.
	Dominant species	Tree, shrub, herb/grass species
	Other notes	
Accession information	Field taxon ID	Scientific name
	Propagule collected	Seed, vegetative cutting, root, rhizome, stolon
	Population distribution	Patchy, uniform

TABLE 3.1 *(continued)*

Data Field	Character	Example
Accession information *(cont)*	Population abundance	Abundant, frequent, occasional, rare
	Plant growth habit	Prostrate, erect, spreading
	Collection source	Wild, cultivated
	Rhizobia collected	Sample number, sample details (nodules, root, soil, etc.)
	Herbarium specimens	Number taken
	Photograph identifier	Film and shot number

Source: Summarized from Steiner and Greene (1996).

FIGURE 3.1. The Preprinted Field Book Used by the Australian Tropical Crops and Forages Collection to Record Data when Collecting Plant Genetic Resources (*Note:* The ATCGRC is the field collection number, while the handwritten AusTRC number (now recorded as AusTRCF) is the accession number given to the collection once it is entered into the AusPGRIS database. This page describes a *Cajanus* collection from the Northern Territory in May 2002.)

often shatter or dislodge from the plant very easily, so it is a good idea to collect a small number of seeds into a small bag (with label) and attach it to the specimen. Detailed instructions on preparing voucher specimens can be found on the Australian Virtual Herbarium Web site (www.chah.gov.au/avh/).

Voucher specimens should be emptied out of the field press each day and placed into a larger press. The newspaper between each specimen should be changed every few days (especially for more succulent species) to prevent mold buildup. Once sufficient specimens have been collected, they can be placed over a low heat to speed up the drying process. It is important to store dry and fresh specimens in separate presses.

On return from the field trip, it is essential to lodge the collected specimens with the state or territory herbarium where the collection took place. The herbarium will then contact researchers once it has identified or verified the taxonomy of each individual specimen. The ATCFC also lodges a duplicate voucher specimen of each Australian collection with a second herbarium.

POSTCOLLECTION ACTIVITIES
AND SEED PROCESSING
FOR EX SITU CONSERVATION

After the collecting mission is finished and you have returned to base, the seed collections must be processed quickly to prevent the loss of seed viability. The seed collections must first be fumigated for insects before they are put in the drying room in the ATCFC. Following fumigation, each collection is cleaned of any plant debris, and seed numbers are determined. The field data are collated with seed numbers, and each collection is given an AusTRCF (Australian Tropical Crops and Forages) accession number. The data are then entered into the National Australian Plant Genetic Resource Information System (AusPGRIS) (www.dpi.qld.gov.au/auspgris) database.

The collections are dried to a seed moisture content of approximately 6 percent in a dehumidified room and then packaged and sealed into laminated foil packets labeled with their AusTRCF number, field collection number, verified taxonomy, and year of collection. The long-term storage facility at the ATCFC operates at −20°C, with seed monitored for decreases in viability during storage. A small number of seeds (around 50) from most accessions are freely

available to bona fide researchers upon request (subject to any distribution restrictions). Seed can be requested either through the
AusPGRIS Web site or by contacting the ATCFC at Locked Mail Bag
1, Biloela QLD 4715, Australia.

RECENT COLLECTING BY THE AUSTRALIAN TROPICAL CROPS AND FORAGES COLLECTION

By following the planning and sampling strategies outlined in this
chapter, the ATCFC undertakes regular collecting trips targeting wild
species of tropical crops and forage species in Australia. Over the
past decade, successful trips have been made the subtropical and
tropical regions of northern New South Wales, Queensland, the
Northern Territory, and Western Australia. Species were prioritized
on filling gaps in the existing ex situ collection at Biloela, based on
geographic range, and also on previously uncollected species. Table 3.2
details the regions visited on each trip, the number of collections
made, and the number of different species conserved.

TABLE 3.2. Australian Tropical Crops and Forages Team Collections, 1992
to 2002

Year	Location	Team Leader	Total Genera	Total Species	Total Seed Collections
2002	NT, WA	S. Dillon	3	11	37
2000	West QLD	S. Dillon	7	15	25
1998	North QLD	S. Dillon	3	6	19
1997/98	NSW, QLD	M. Rettke	2	3	26
1997	NT	I. Cowie	2	6	7
1996	NT, WA	J. Corfield	3	15	71
1995	QLD, NT	P. Lawrence	2	13	44
1994	QLD, NT, WA	P. Lawrence	2	19	172
1992	NT	P. Lawrence	2	5	14

Note: NT = Northern Territory; WA = Western Australia; QLD = Queensland;
NSW = New South Wales.

CONCLUDING REMARKS

Adopting the strategies outlined in this chapter will enable plant collectors to confidently plan a collecting mission to meet their collection objectives. This will ensure that the germplasm collected is representative of the genetic diversity and the geographic distribution of plant species and contains viable seed for long-term conservation in ex situ genetic resource centers.

REFERENCES

Baskin, C.C. and Baskin, J.M. (2001). *Seeds: Ecology, Biogeography, and Evolution of Dormancy and Germination.* San Diego: Academic Press.

Brown, A.D.H. and Marshall, D.R. (1995). A basic sampling strategy: Theory and practice. In: Guarino, L., Rao, V.R., and Reid, R. (eds.), *Collecting Plant Genetic Diversity: Technical Guidelines.* Wallingford, UK: CAB International, pp. 75-91.

Chapman, C.G.D. (1989). Collection strategies for the wild relatives of field crops. In: Brown, A.H.D., Marshall, D.R., Frankel, O.H., and Williams, J.T. (eds.) *The Use of Plant Genetic Resources.* Melbourne: Cambridge University Press, pp. 265-279.

Date, R.A. (1995). Collecting *Rhizobium, Frania* and mycorrhizal fungi. In: Guarino, L., Rao, V.R., and Reid, R. (eds.), *Collecting Plant Genetic Diversity: Technical Guidelines.* Wallingford, UK: CAB International, pp. 551-560.

Engelmann, F. and Engels, J.M.M. (2002). Technologies and strategies for ex situ conservation. In: Engels, J.M.M., Rao, V.R., Brown, A.H.D., and Jackson, M.T. (eds.), *Managing Plant Genetic Diversity.* New York: CABI Publishing, pp. 89-103.

Engels, J.M.M., Arora, R.K., and Guarino, L. (1995). An introduction to plant germplasm exploration and collecting: Planning, methods and procedures, follow-up. In: Guarino, L., Rao, V.R., and Reid, R. (eds.), *Collecting Plant Genetic Diversity: Technical Guidelines.* Wallingford, UK: CAB International, pp. 31-63.

Ganeshan, S. and Rajashekaran, R.K. (2000). Current status of pollen cryopreservation research: Relevance to tropical horticulture. In: Engelmann, F. and Takagi, H. (eds.), *Cryopreservation of Tropical Plant Germplasm: Current Research Progress and Applications.* Rome: Japan International Centre for Agricultural Sciences, Tsukuba/International Plant Genetic Resources Institute, pp. 360-365.

Guarino, L. (1995). Assessing the threat of genetic erosion. In: Guarino, L., Rao, V.R., and Reid, R. (eds.), *Collecting Plant Genetic Diversity: Technical Guidelines.* Wallingford, UK: CAB International, pp. 67-74.

Harlan, J.R. and de Wet, J.M.J. (1971). Toward a rational classification of cultivated plants. *Taxon* 20, 509-517.

Hoyt, E. (1988). *Conserving the Wild Relatives of Crops.* Rome: IBPGR.

Ishii, T., Brar, D.S., Multani, D.S., and Khush, G.S. (1994). Molecular tagging of genes for brown planthopper resistance and earliness introgressed from *Oryza australiensis* into cultivated rice, *O. sativa. Genome* 37, 217-221.

Ladizinsky, G. (1989). Ecological and genetic considerations in collecting and using wild relatives. In: Brown, A.H.D., Marshall, D.R., Frankel, O.H., and Williams, J.T. (eds.), *The Use of Plant Genetic Resources.* Melbourne: Cambridge University Press, pp. 297-305.

Loveless, M.D. and Hamrick, J.L. (1984). Ecological determinant of genetic structure in plant populations. *Annual Review of Ecology and Systematics* 15, 65-95.

Marshall, D.R. and Broué, P. (1981). The wild relatives of crop plants indigenous to Australia and their use in plant breeding. *Journal of the Australian Institute of Agricultural Science* 47, 149-154.

Marshall, D.R. and Brown, A.H.D. (1975). Optimum sampling strategies in genetic conservation. In: Frankel, O.H. and Hawkes, J.G. (eds.), *Genetic Resources for Today and Tomorrow.* Cambridge: Cambridge University Press, pp. 53-80.

Marshall, D.R. and Brown, A.H.D. (1983). Theory of forage plant collection. In: McIvor, J.G. and Bray, R.A. (eds.), *Genetic Resources of Forage Plants.* Melbourne: CSIRO, pp. 135-138.

Namkoong, G. (1988). Sampling for germplasm collections. *HortScience* 23, 79-81.

Panis, B., Schoofs, H., Thinh, N.T., and Swennen, R. (2000). Cryopreservation of proliferation meristem cultures of banana. In: Engelmann, F. and Takagi, H. (eds.), *Cryopreservation of Tropical Plant Germplasm: Current Research Progress and Applications.* Rome: Japan International Centre for Agricultural Sciences, Tsukuba/International Plant Genetic Resources Institute, pp. 238-244.

Rao, N.K., Reddy, L.J., and Bramel, P.J. (2003). Potential of wild species for genetic enhancement of some semi-arid food crops. *Genetic Resources and Crop Evolution* 50, 707-721.

Reddy, L.J., Saxena, K.B., Jain, K.C., Singh, U., Green, J.M., Sharma, D., Faris, D.G., Rao, A.N., Kumar, R.V., and Nene, Y.L. (1997). Registration of high-protein pigeonpea elite germplasm ICPL 87162. *Crop Science* 37, 294.

Sahijram, L. and Rajasekharan, P.E. (1997). Tissue culture strategies applicable to in vitro conservation of tropical fruit crops. In: Rao, V.R. and Arora, R.K. (eds.), *Proceedings of Tropical Fruits in Asia: Diversity, Maintenance, Conservation and Use.* Bangalore: IPGRI, pp. 113-119.

Schäfer-Menuhr, A., Schumacher, H.M., and Mix-Wagner, G. (1997). Cryopreservation of potato cultivars—design of a method for routine application in genebanks. *Acta Horticulturae* 447, 477-482.

Steiner, J.J. and Greene, S.L. (1996). Proposed ecological descriptors and their utility for plant germplasm collections. *Crop Science* 36, 439-451.

UNCED (1992). Convention on biological diversity. *United Nations Conference on Environment and Development.* Geneva: United Nations.

von Bothmer, R. and Seberg, O. (1995). Strategies for the collecting of wild species. In: Guarino, L., Rao, V.R., and Reid, R. (eds.), *Collecting Plant Genetic Diversity: Technical Guidelines*. Wallingford, UK: CAB International, pp. 93-111.

Weir, B.S. (1990). *Genetic Data Analysis*. Sunderland: Sinauer Associates.

Xiao, J., Li, J., Grandillo, S., Ahn, S.N., Yuan, L., Tanksley, S.D., and McCouch, S.R. (1998). Identification of trait-improving quantitative trait loci alleles from a wild rice relative, *Oryza rufipogon*. *Genetics* 150, 899-909.

Yonezawa, K. (1985). A definition of the optimal allocation of effort in conservation of plant genetic resources, with application to sample size determination for field collection. *Euphytica* 34, 345-354.

Yonezawa, K. and Ichihashi, H. (1989). Sample size for collecting germplasm from natural plant populations in view of the geographic multiplicity of seed embryos borne on a single plant. *Euphytica* 41, 91-97.

Chapter 4

The Role of Genetic Resources
Held in Seed Banks
in Plant Improvement

Tsukasa Nagamine

INTRODUCTION

There are more than 1,300 national and regional germplasm collections, of which 397 are maintained under medium- or long-term storage conditions, besides the gene banks of International Agricultural Research Institutes (IARI) around the world, according to the *Report on the State of the World's Plant Genetic Resources* (International Plant Genetic Resources Institute [IPGRI], 1996) presented at the International Technical Conference on Plant Genetic Resources held in Leipzig, Germany, in 1996. The establishment, scale, and capacities of the gene banks as well as the kinds of genetic resources conserved vary.

In this chapter *gene bank* is defined as a facility where plant seeds and vegetative propagated organs are preserved in appropriate storage conditions. An *in vitro storage facility* is defined as one that includes germplasm preservation of vegetative organs such as fruit trees in the field, and a *seed bank* is a facility where only seeds are preserved.

The technical development of medium- and long-term storage of seeds under low-temperature and low-humidity conditions and the construction of the necessary facilities for seed preservation at low cost are the main developments that allowed the establishment of many gene banks in the world. Studies on seed preservation (Ito, 1965) and seed longevity (Roberts, 1973), the improvement of air-conditioning apparatuses, the development of insulating materials,

doi:10.1300/5546_04

57

and the progress in seed-drying machines form the fundamental theoretical and technical bases for the long-term storage of seed materials. The International Board for Plant Genetic Resources (IBPGR) (1985) proposed a guideline for establishment of gene banks.

Various types of plant genetic resources, namely, local and improved cultivars of crops, related wild species, and experimental lines, are now preserved in gene banks and have been used for crop breeding and for research purposes. Crops that have been established and maintained by local farmers for a long period of time and or developed by plant breeders are integrated living accumulators composed of numerous selected genes, and they themselves may be considered gene banks harboring precious genes (Koyama, 1984).

The purpose of gene banks is to preserve genetic resources that will sooner or later become extinct and to use them for future crop breeding and/or scientific studies. To use genetic resources effectively, it is important to provide an adequate number of authentic plant genetic resources and their related information showing such passport data as origin, morphological and ecological characters, biotic and abiotic stresses, and other agronomic traits and to respond promptly to users' demands.

Among the plant genetic resources preserved ex situ in gene banks, local cultivars are a major part of the collection and are expected to have numerous useful genes and/or alleles. Therefore, in this chapter, we focus on crop landraces or local cultivars that have been maintained and conserved by local farmers in their fields for long periods of time until recently and that are now preserved in seed banks, natural genes and alleles newly found in local cultivars, and the development of novel evaluation methods. The significance and importance of the preservation of local cultivars in seed banks is emphasized. Finally, new technical developments for seed conservation in genebanks are proposed. The following examples illustrate the role of seed banks.

NEW ALLELES ON Wx GENES
FOUND IN LOCAL WHEAT CULTIVARS

To promote the use of plant genetic resources preserved in seed banks, the evaluation of several characters of biotic/abiotic stresses and physiochemical properties of plant organs harvested is the most significant task for gene bank managers. Invention or modification of new evaluation methods for each character is very important for future use.

Various types of mutants in endosperm starch have been reported for maize, rice, and foxtail millets. Glutinous, nonglutinous, and dull endosperm types with low amylose content were spontaneously found in local cultivars of rice and foxtail millets, and artificial mutants were also created by radiation treatment and by chemical mutagen treatments.

Thus far, only nonglutinous endosperm has been known to exist in common wheat germplasm worldwide. Common wheat has an allohexaploid genome constitution of AABBDD; thus all geneticists and breeders had believed that it was difficult to improve the starch properties of wheat endosperm through genetic manipulations such as artificial mutations. As a result, the study of genetic improvement of the endosperm starch of wheat started later than that of maize and rice.

Most of the wheat grains harvested in Japan are used as the flour materials for udon noodles, of which Japanese consumers prefer the sticky type. Amylose content in the endosperm is closely related to the stickiness of the noodles (Oda et al., 1980), and this type of udon noodle can be made by lowering the amylose content of wheat endosperm.

The amylose content of endosperm is regulated by the amount of Wx proteins produced (Yamamori and Nakamura, 1994). The amount of Wx protein expression can be assayed by electrophoresis. However, because common wheat has a hexaploid genome constitution, it is very difficult to estimate the separate expressions of the three Wx proteins Wx-A, Wx-B, and Wx-D using one-dimensional electrophoresis. Thus a two-dimensional (2D) electrophoresis method was modified to include polyacrylamide electrophoresis at the first dimension and isoelectric focusing at the second dimension. Using null-tetrasomic series of the common wheat cultivar (cv.) Chinese Spring, Wx protein spots on the 2D gel were analyzed. The relative positions on the gel for the three Wx proteins were determined and the chromosomal locations of *Wx-A1, Wx-B1,* and *Wx-D1* loci were proved to be on chromosomes 7A, 4A, and 7D, respectively (Nakamura et al., 1993).

Two wheat breeding lines, 'Komugi Kanto 79' and 'Komugi Kanto 107', bred in Japan have a rather low amount of amylose content of 22 percent, and the amount of Wx protein expression was also low. According to the results on the amount of Wx proteins analyzed by 2D electrophoresis, both lines have null alleles on the Wx-A1 and Wx-B1 loci (Wx-A1b, Wx-B1b), and Wx proteins are hardly produced at these two loci (Nakamura et al., 1993). Therefore, if another null allele on the Wx-D1 locus could be found in wheat genetic resources, three kinds of spontaneous deficient mutations on the Wx loci could be prepared, making it highly possible to create a glutinous-endo-sperm type wheat by artificial cross hybridization between the above-mentioned mutants that do not produce any amylose.

A large-scale screening was conducted to detect spontaneous defi-cient mutant cultivars of the Wx-D1 locus using worldwide wheat germplasm preserved in the gene bank of the National Institute of Agrobiological Sciences (NIAS) (Yamamori et al., 1994). After the screening of 1,960 cultivars, the Chinese local cultivar Bai Huo, which has a deficient allele of the Wx-D1 locus, Wx-D1b, was finally found (Table 4.1). Three kinds of deficient alleles on the Wx-A1, Wx-B1, and Wx-D1 loci were found to create a new genotype of glutinous-endosperm wheat. By artificial cross hybridization between 'Komugi Kanto 107' and 'Bai Huo', glutinous wheat cultivars Mochiotome and Hatsumochi were bred and released (Yamamori and Nakamura, 1994). These results prove that genetic modification of the endo-sperm starch of common wheat is possible.

'Bai Huo' was introduced from China and placed in the NIAS gene bank in 1974, and its morphological characters and biotic and abiotic characters have been described and evaluated. However, this local cultivar was not distributed to any users until it was proved to have a rare allele, the null allele of Wx-D1 (Wx-D1b). It then became the breeding material for creating a glutinous wheat cultivar in 1992. The research results on the modification of 2D electrophoresis that can discriminate the alleles at the Wx loci by genome and the large-scale screening for a new deficient allele of Wx-D1b on the Wx-D1 locus demonstrated the importance of germplasm conservation in the gene bank. Unless 'Bai Huo' was preserved in the NIAS gene bank, the glutinous-type wheat cultivars would never have been bred. Safe and steady conservation of germplasm is the most important task of the gene bank.

TABLE 4.1. Distribution of Wheat Cultivars Lacking the Wx-A1, Wx-B1, and Wx-D1 Proteins As Revealed by Modified SDS-PAGE and 2D-PAGE

Origin	Number of Cultivars Examined	Missing Wx Protein		
		Wx-A1	Wx-B1	Wx-D1
Japan	462	75	16	0
South and North Korea	93	10	1	0
China	308	3	12	1
India	50	3	25	0
Pakistan	85	0	13	0
Afghanistan	59	0	13	0
Turkey	156	81	0	0
Australia	127	1	51	0
North America	315	3	19	0
Western Europe	172	1	4	0
Russia	133	0	5	0
Total	1,960	177	159	1

Source: Adapted from Yamamori et al. (1994).

NULL ALLELE OF LIPOXYGENASE-3
FROM RICE LOCAL CULTIVARS
USING CORE COLLECTION

Besides the safe preservation of certain amounts of collections, the provision of novel and unique genetic resources to users is the issue that gene bank managers are the most concerned about. A core collection is one of the unique genetic resources provided by gene banks.

The significance of a core collection, which is composed of a small number of accessions and is presumed to represent the genetic variations of crop genetic resources of over several tens of thousands in total that are conserved in the gene bank, has been reviewed (Hodgkin et al., 1995). When we need to screen several characters effectively with limited time and cost, a core collection can be useful to plant breeders and researchers. According to the definition by Frankel and Brown (1984), "A core collection consists of a limited set of accessions derived from an existing germplasm collection, chosen to represent the genetic spectrum in the whole collection. The core should include as much as possible of the genetic diversity." The International Rice Research Institute proposed a core collection of rice in 2001, and the worldwide core collection of barley (Table 4.2) was established by international collaboration under the guidance of a barley core collection coordinating committee (Knupffer and Van Hintum, 2003).

One of the purposes of using a core collection established in a gene bank is to preliminarily survey the magnitude of genetic variations of a target character. The core collection is intended to be an effective tool for prescreening of a new character. Plant breeders are always seeking novel genes or alleles, by inventing evaluation methods, that are useful for crop breeding. Many of the important genes deployed in plant breeding so far seem to have been very rare before their use. Studying core collections to find rare alleles is not always effective. However, we describe a successful example.

TABLE 4.2. Tentative Sizes of Regional Subsets of the Cultivars and Landrace Groups of the Barley Core Collection

Subgroup	Cultivars	Landraces
West Asia, North Africa	15	300
South and East Asia	80	300
Ethiopia	5	100
Europe	200	80
North and South America	150	30
Australia, New Zealand, South Africa, and other regions	35	0

Source: Adapted from Knupffer and Van Hintum (2003).

Notes: The barley gene pool was initially divided into five main categories, namely, (1) cultivars (500 accessions), (2) landraces (800), (3) *Hordeum vulgare* ssp. *spontaneum* (150-200), (4) other *Hordeum* wild species (60-100, two per species), and (5) genetic stocks and reference material (max. 200). The further subdivision of the cultivated barleys and ssp. *spontaneum* follows ecogeographical criteria, whereas the wild species of *Hordeum* are divided into taxonomical and ecogeographical criteria. In this table "cultivars" mean improved cultivars bred by breeders, while "landraces" mean local cultivars developed by farmers.

Deterioration of the rice grains and development of a stale flavor during the storage of brown rice is a serious problem that reduces the quality of the stored rice. Degradation of lipids is responsible for these changes, and lipoxygenase (LOX) catalyzes the peroxidation of linoleic and linolenic acids to form hydroperoxides (Yasumatsu and Moritaka, 1964; Suzuki et al., 1993). Hydroperoxides are transformed into volatile compounds. Three isozymes of lipoxygenase are recognized and the isozyme *Lox-3* is the major one in rice grains; it represents 80 to 90 percent of lypoxygenase (Ida et al., 1983).

To breed a new rice variety that will not develop a stale flavor during storage, a survey for the null allele at the *Lox-3* locus was started using rice genetic resources. More than 31,000 rice accessions are now preserved in the NIAS gene bank, but, it is difficult to find a targeted cultivar with a null allele at the *Lox-3* locus among them. A small-scale core collection including only 93 cultivars, mostly of local cultivars, originating from all over the world was used as the material for a preliminary screening.

Screening for a cultivar harboring a null allele on the *Lox-3* locus using a monoclonal antibody identified the local cultivar Daw Dam, originating from Thailand, to have a null allele (Suzuki et al., 1993). This cultivar has a glutinous endosperm, but lypoxygenase is not genetically linked to this character. This is a successful example of finding a rare null allele on the *Lox-3* locus using the core collection comprising a small number of cultivars. After this preliminary screening, a second survey to detect other cultivars with a null allele on *Lox-3* was initiated using 471 randomly chosen local cultivars originating from 13 Asian countries. Cultivars with a null allele of *Lox-3* were found only among those from Southern China, North Vietnam, Laos, Thailand, and Myanmar (Table 4.3; Suzuki et al., 2000). The occurrence of the null allele at *Lox-3* was 4.5 percent, which is not a very low frequency. Using a small-scale core collection of 93 local cultivars, accessions with a null allele on *Lox-3* might be found easily.

The result demonstrates the effectiveness of a core collection in prescreening for a new character. Therefore, it is recommended that national gene banks, which preserve unique genetic resources originating from their own countries, establish and provide core collections of indigenous crops.

TABLE 4.3. Distribution of LOX-3 Deficient Cultivars of Rice in Asian Countries

Countries	Number of Cultivars Examined	LOX-3 Deficiency	Percentage of Occurrence
Japan	17	0	0.0
Philippines	31	0	0.0
Indonesia	32	0	0.0
China	86	7	8.1
Vietnam	40	1	2.5
Laos	61	6	9.8
Thailand	28	1	3.6
Myanmar	51	6	11.8
Malaysia	15	0	0.0
Bhutan	20	0	0.0
Nepal	15	0	0.0
India	59	0	0.0
Pakistan	16	0	0.0
Total	471	21	4.5

Source: Suzuki et al. (2000).

CHARACTERIZING NATURAL GENES CONTROLLING THE HEADING OF RICE BY QTL ANALYSIS

The heading date or flowering time of crops is among the most important genetically controlled characters for crop production. Timing of the heading date of crops at each local area is very much related to the local adaptability of crops, and this character is the most essential character in crop breeding. Classic genetic analysis of the heading date of sorghum, wheat, barley, and maize has been carried out worldwide, and many kinds of genes controlling the character have been reported.

A photoperiod-sensitive locus *Se-1* was reported in rice plants (Chang et al., 1969; Yokoo et al., 1980), and other related loci have also been reported. It is understood that the heading date of rice plants is controlled by the duration and degree of the basic vegetative phase and of the photoperiod sensitivity, and many genes are known to contribute to these two characters. Heading characters of rice plants are controlled not only by a major gene but also by minor genes. The determination of chromosomal localization of minor genes related to the heading date is very difficult to establish with traditional genetic analyses, because they only provide information on the number of regulating genes and their quantitative functions. Therefore, there are technical limitations to traditional genetic analysis and successive breeding. The heading date of rice plants is regulated by major qualitative genes such as the *Se-1* locus, which responds to the short day duration, and other minor quantitative genes, which are coordinately regulating the heading date of crops.

Genetic analysis and identification of the genes controlling agronomic characters have been undertaken using molecular markers in plant species (Tanksley, 1993). Among many natural genes regulating agronomic characters, the genes controlling the heading of rice plants have been analyzed. To investigate the heading performance of rice plants in detail, using F2 hybrid populations bred between the *indica* cv. Kasalath and the *japonica* cv. Nipponbare, genetic analysis of heading date was conducted by quantitative trait loci (QTL) analysis. The *Hd1* to *Hd5* loci were mapped based on the analysis of the F2 hybrid population (Yano et al., 1997). The *Hd-7, Hd-8,* and *Hd-11* loci were found by using BC1F5 lines (Lin et al., 2002), and other loci, *Hd-6, Hd-9, Hd-10, Hd-12, Hd-13,* and *Hd-14,* were detected by using advanced backcross progeny, such as BC3F2 or BC4F2. The number of loci controlling the heading date is rather greater than was previously expected. Among them, the QTL locus, *Hd-1,* which is located on chromosome 6, is an ortholog of the *CONSTANS (CO)* gene of *Arabidopsis* plants and allelic to the *Se-1* locus. QTL *Hd-6,* located on chromosome 3, is involved in photoperiod sensitivity. *Hd-6* encodes the α-subunit of protein kinase CK2 (Yano et al., 2001).

These studies reveal the nature of heading traits in rice cultivars, and new natural genes controlling heading traits are found with new evaluation methods. New natural genes that have been accumulated in local cultivars may be explored by QTL analysis. These cultivars have been grown on farmers' fields and maintained and conserved by local farmers.

DISCOVERY OF THE GENE
CONTROLLING SEED LONGEVITY OF RICE

Seed longevities, of plant genetic resources vary according to environmental conditions such as relative humidity and temperature of the storage facility and the water content of seeds. Generally, longer seed longevity can be achieved by storing seeds in conditions of low atmospheric moisture content and low temperature, and using seeds with a low water content. Practically, many seed banks adopt a set of temperature, seed moisture content, and humidity conditions because little difference in the optimum conditions needed by crop genotypes. Studies on the seed longevity of wheat (Piech and Supryn, 1979) and soybeans (Singh and Ram, 1986; Verma and Ram, 1987) have been conducted, but the mode of inheritance of longevity was not obvious.

A QTL study of loci regulating longevity of rice seeds was undertaken using 98 recombinant inbred lines (RIL) of the *japonica* cv. Nipponbare and the *indica* cv. Kasalath. After harvest, seeds from 98 RILs were stored in a room at 30°C for 12 months to lose seed longevity. After 12 months, an accelerated aging treatment was given to the seeds for two months at 30°C and a relative humidity of 15 percent. Then, the germination ratio was measured for each RIL at 20°C (night) to 30°C (daytime with light), and QTL analysis was conducted to determine genetic control of the germination rates.

As a result, three QTLs, *qLG-2, qLG-4,* and *qLG-9,* were detected on chromosomes 2, 4, and 9 (Table 4.4 and Figure 4.1). Among them, the QTL locus *qLG-9* on chromosome 9 explained 60 percent of the total phenotypic variations as the major gene. It was revealed that all the alleles at the QTL loci from the *indica* cv. Kasalath prolong seed longevity, but those from the *japonica* cv. Nipponbare do not.

Detection of the three QTLs concerned with seed longevity using *indica* and *japonica* hybrid experimental lines revealed the nature of seed longevity in rice. This finding contradicted the earlier assumption that there was no genetic variation in seed longevity. Because there is a large intraspecific difference in seed longevity in rice, we had to investigate the seed germination of every accession preserved in gene banks and adjust the interval of seed multiplication for each accession, even if they were entered into the gene banks at the same time. The monitoring of seed germination by random sampling of a

TABLE 4.4. Putative QTLs for Seed Dormancy and Seed Longevity in Rice

Trait	QTL	Nearest Marker Locus	Putative QTLs	Chromosome GLM/SAS Probability
Seed dormancy	qSD-1	R1613	1	0.0006
	qSD-3	C25	3	0.0014
	qSD-5	R1838	5	0.0001
	qSD-7	R1357	7	0.0076
	qSD-11	C189	11	0.0018
Seed longevity	qLG-2	C1470	2	0.0007
	qLG-4	R514	4	0.0012
	qLG-9	R79	9	<0.0001

Source: Adapted from Miura et al. (2002).

small number of accessions is not sufficient for the exact determination of seed longevity.

The three QTLs, *qLG-2, qLG-4,* and *qLG-9,* controlling seed longevity are not located on the same chromosomes as the five QTLs, *qSD-1, qSD-3, qSD-5, qSD-7, qSD-11,* regulating seed dormancy. (Table 4.4; Miura et al., 2002). It had seemed that seed dormancy and seed longevity were related to each other; however, based on the QTL analysis, the mechanisms of the two phenomena are different, and they are genetically regulated independently. This research gave us a new direction for the management of the seeds of plant genetic resources in gene banks. Seed dormancy is a practical problem for seed bank preservation. Dormant seeds sometimes show a low germination ratio in the germination test of the initial seed preservation and in the monitoring of seed germination after storage, and the germination ratio due to longevity is then not exact. Thus the seed germination ratio was investigated practically in the seed bank after dormancy was broken by heat shock treatment and chemical treatment with 0.2 percent nitric acid or gibberellic acid. Expressed sequence tag markers for the QTL genes controlling seed longevity will help detect the presence of seed dormancy in each accession.

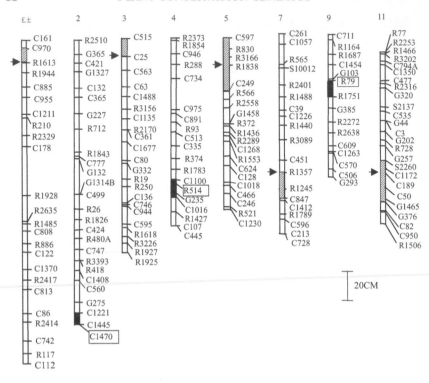

FIGURE 4.1. Chromosomal locations of QTLs for primary dormancy and seed longevity in rice. Striped and black bars represent putative region of QTLs for primary dormancy and seed longevity, respectively. A reduction of 0.5 LOD value from LOD peaks was used to define left and right borders of the confidence interval in MAPMAKER/QTL. Arrows and boxes indicate the nearest marker locus to the QTLs for primary dormancy and seed longevity, respectively, in markers that were significant at the 0.01 probability level based on ANOVA. Chromosomes with no QTL are omitted in this figure.

EFFECTIVE MANAGEMENT OF THE STOCK DATABASE BY SEED LOT UNIT

According to the *Report on the State of the World's Plant Genetic Resources* (IPGRI, 1996), 77 countries have seed storage facilities suitable for medium- and long-term storage, but fewer than half can offer secure, long-term management of accessions. Several gene banks, including

IARIs, have duplicate storage systems for seeds of plant genetic resources to secure the preservation of authentic seeds. Seeds acquired on-farm or in situ are divided, and half are put into medium-term storage and the rest into long-term storage. This system was developed by Ito (1965). The fundamental idea is to preserve seeds without advancing any generations of seeds in gene banks. IPGRI (1985) recommended preservation conditions for medium- and long-term storages to ensure that the viability of accessions remains above 65 percent for 10 to 20 years, and if the germination ratio of any accessions declines below 85 percent of the initial value, fresh seeds of the accessions should be multiplied by using the original seeds preserved in long-term storage. If the amounts of seed preserved in medium-term storage become too small, fresh seeds should be produced. After these seed multiplications, seeds of a new generation of the same accession are preserved separately in the seed bank.

To manage the stock database in medium- and long-term storages, while discriminating each generation of the original and multiplicated seeds, the concept of seed lot units for the same accessions is useful. Each seed lot unit has information on the number of generations after storage in the gene bank, the year and location of seed multiplication, the stock number at the seed storage facilities, and the amount of seed preserved. Generally, a lot unit of old distribution seeds is preserved in the medium-term storage in addition to the fresh distribution seeds in gene banks. Some gene banks apply the concept of "multiplication year" to different seed lots of the same accessions. However, not only the multiplication year but also other information can be dealt with in the seed lot unit concept, and it is advisable to use seed lot units in stock database management in gene banks.

CONSERVATION AND MULTIPLICATION OF PLANT GENETIC RESOURCES IN THE COUNTRY OF ORIGIN

More than 6 million accessions of plant genetic resources are preserved in gene banks worldwide (IPGRI, 1996). The most difficult management task is seed multiplication of the resources collected and introduced from foreign countries or areas, particularly from different climatic regions. Multiplication of seeds that originated in a temperate region is easily done in the gene banks located in a temperate region with the same climatic conditions; however, multiplication of seeds from subtropical and tropical regions is very difficult at gene banks in temperate regions. In particular, most of the genetic resources, namely, rice, legumes of the *Vigna* species, amaranth, sorghum, and finger millets, that originate in tropical and subtropical regions have a strong photoperiod sensitivity or a long vegetative growth. They do not head or flower under natural day-length conditions in a temperate region, and the number of accessions that head or flower and are harvested in the natural climatic conditions of the region is limited. Thus, for photoperiod-sensitive rice and legumes in the region of Japan, the NIAS gene bank conducts seed multiplication in artificial day-length-controlling facilities. Because of the limited facility area, however, the number of plants per accession of seed multiplication is limited.

Genetic bottlenecks or genetic drift (Singh and Williams, 1984) can happen during routine seed multiplication, and genetic drift can occur in the multiplicated seed population. Ex situ collection or preservation itself can induce genetic bottlenecks or genetic drift in the resources introduced into gene banks from on-farm and in situ collections. It is also difficult to rejuvenate the seeds of thermosensitive accessions from high-latitude areas in tropical and subtropical areas.

For those gene banks that work together internationally, it is advised that the conservation, multiplication, and distribution of each accession of the country of origin should be conducted by each gene bank. At the Nordic Gene Bank, species used in Nordic agriculture and horticulture and their wild relatives are preserved. In addition, the species of current interest to biotechnology, as well as landscape plants, medicinal plants, culinary herbs, and plants with industrial uses, are being considered for preservation. Species that are cultivated elsewhere and are found in the wild in Nordic countries are also considered. The species within this scope are called *mandate species* (Vetelainen and Endresen, 2000).

Recently the NIAS genebank started collaborative seed multiplication between subtropical and tropical countries utilizing their environmental conditions. This is one of the solutions for the seed multiplication of photoperiod-sensitive cultivars.

EVALUATION DATABASE FOR HETEROGENEOUS PLANT GENETIC RESOURCES

Many gene banks now conduct characterization and evaluation studies according to the descriptors of each crop species published by IPGRI or established by each institute. Until recently, field notes were the main source of documentation data. Now computer databases more effectively deal with the huge amount of information that is routinely generated. For example, the characterization and evaluation data of an improved cultivar and/or mutant line are normally reported as one numerical value or score due to genetic homogeneity. However, many local cultivars generally comprise different genotypes in one accession, and they show heterogeneous morphological statures and different responses to biotic/abiotic stresses.

As mentioned previously, the local cv. Daw Dam was found to have a null allele on the *Lox-3* locus. However, when it was distributed to the NIAS gene bank, this cultivar was revealed to be heterogeneous for this locus. The allele on the *Lox-3* locus of the cv. Daw Dam accession was then investigated individually by electrophoresis. Two types of cv. Daw Dam, one with and one without a null allele on the *Lox-3* locus, are now preserved in the gene bank.

When we used field notes to fill in these heterogeneous data, there were no practical problems, and documentation of heterogeneous accessions in the electronic database is not always easy. In order to express heterogeneous data for accessions in gene bank databases, the databases need to be improved.

CONCLUSION

Local cultivars or landraces established and maintained by local farmers in their fields over a long period of time contain useful natural genes. When they are preserved in gene banks they will play an important role as providers of natural genes and alleles to plant breeders and crop and plant scientists in the future. Because natural genes in local cultivars were induced by spontaneous mutations and have been maintained on farmers' fields for a long time, it is expected that they can easily be utilized under severe natural conditions. Whether this happens all depends on the continuing improvement of evaluation methods, particularly to screen novel genotypes from the accessions preserved in gene banks. The safe preservation of genetic resources and the development of new methods for the evaluation and provision of novel natural genes from local cultivars are important task of gene banks.

REFERENCES

Chang, T.T., Li, C.C., and Vergara, B.S. (1969). Component analysis of duration from seeding to heading in rice by the base vegetative phase and photoperiod-sensitive phase. *Euphytica* 18, 79-91.

Frankel, O.H. and Brown, A.F.D. (1984). Plant genetic resources today: A critical appraisal. In: Holden, J.H.W. and Williams, J.T. (eds.), *Crop Genetic Resources: Conservation and Evaluation.* Allen & Unwin, London, pp. 247-257.

Hodgkin, T., Brown, A.H.D., van Hintum, T.J.L., and Morales, E.A.V. 1995. *Core collection of plant genetic resources.* New York: John Wiley & Sons.

Ida, S., Masaki, Y., and Morita, Y. (1983). The isolation of multiple forms and product specificity of rice lipoxygenase. *Agriculture Biology and Chemistry.* 47, 637-641.

International Board for Plant Genetic Resources (1985). *Handbooks for Genebanks, Nos. 1 and 2.* IBPGR, Rome, Italy, pp. 210, 667.

International Plant Genetic Resources Institute (1996). *Report on the State of the World's Plant Genetic Resources.* IPGRI, Rome, Italy.

Ito, H. (1965). Studies on maintenance of genetic stocks and a breeding system for rice plants based on long-term seed storage [Japanese with English summary]. *Bulletin of the National Institute of Agricultural Science* D13, 163-230.

Knupffer, H. and Van Hintum, T. (2003) Summarized diversity—the barley core collection. In: von Bothmar, R., Van Hintum, T., Knupffer, H., and Sato, K. (eds.), *Diversity of Barley* (Hordum vulgare), Elsevier Science, Amsterdam, the Netherlands, pp. 259-267.

Koyama, T. (1984). *Resource Botany.* Kodansha Scientific, Tokyo, 117. [In Japanese.]

Lin, H., Ashikari, M., Yamanouchi, U., Sasaki, T., and Yano, M. 2002. Identification and characterization of a quantitative trait locus, Hd9, controlling heading date in rice. *Breeding Science* 52, 35-41.

Miura, K., Lin, S.Y., Yano, M., and Nagamine, T. (2002). Mapping quantitative trait loci controlling seed longevity in rice (*Oryza sativa* L.). *Theoretical and Applied Genetics* 104, 981-986.

Nakamura, T., Yamamori, M., Hirano, H., and Hidaka, S. (1993). Identification of three waxy (Wx) proteins in wheat (*Triticum aestivum* L.). *Biochemistry Genetics* 31, 75-86.

Oda, M., Yasuda, Y., Okazaki, S., Yamauchi, Y., and Yokohama, T. (1980). A method of flour quality assessment for Japanese noodles. *Cereal Chemistry* 57, 253-254.

Piech, J. and Supryn, S.T. (1979). Effect of chromosome deficiencies on seed viability in wheat, *Triticum aestivum* L. *Annals of Botany* 43, 115-118.
Roberts, E.F. (1973). Predicting the storage life of seeds. *Seed Science & Technology* 1, 499-514.
Singh R.B. and Williams, J.T. (1984). Maintenance and multiplication of plant genetic resources. In: Holden, J.H.W. and Williams, J.T. (eds.), *Crop Genetic Resources Conservation and Evaluation.* London: Allen & Unwin, pp. 120-130.
Singh, R.K. and Ram, H.H. (1986). Inheritance study of soybean seed storability using an accelerated aging test. *Field Crop Resources* 13, 89-98.
Suzuki, Y., Ise, K., and Nagamine, T. (2000). Geographical variation of the gene, *(lox3[t]),* causing lipoxygenase-3 deficiency in Asian rice varieties. *Rice Genetic Newsletter* 17, 13-14.
Suzuki, Y., Nagamine, T., Kobayashi, A., and Ohtsubo, K. (1993). Detection of new variety lacking lipoxygenase-3 by monoclonal antibodies. *Japanese Journal of Breeding* 43, 405-409.
Tanksley, S.D. (1993). Mapping polygenes. *Annual Review of Genetics* 27, 205-233.
Verma, V.D. and Ram, H.H. (1987). Genetics of seed longevity in soybean. *Crop Improvement* 14, 42-46.
Vetelainen, M. and Endresen, D.T. (2000). Barley core collection at the Nordic Gene Bank. In: IPGRI (ed.), *Report of Working Group on Barley.* IPGRI, Rome, Italy, pp. 35-41.
Yamamori, M. and Nakamura, T. (1994). Production of waxy wheat by genetically eliminating Wx proteins. *Gamma Field Symposia* 33, 63-74.
Yamamori, M., Nakamura, T., Endo, T.R., and Nagamine, T. (1994). Waxy protein deficiency and chromosomal location of coding genes in common wheat. *Theoretical and Applied Genetics* 89, 179-184.
Yano, M., Horushima, Y., Nagamura, Y., Kurata, N., Minobe, Y., and Sasaki, T. 1997. Identification of quantitative trait loci controlling heading date in rice using a high-density linkage map. *Theor. Appl. Genet.* 95, 1025-1032.
Yano, M., Kojima, S., Takahashi, Y., Lin, H., and Sasaki, T. (2001). Genetic control of flowering time in rice, a short-day plant. *Plant Physiology* 127, 1425-1429.
Yasumatsu, K. and Moritaka, S. (1964). Fatty acid compositions of rice lipid and their changes during storage. *Agriculture Biology and Chemisty* 28, 257-264.
Yokoo, M., Kikuchi, F., Nakane, A., and Fujimaki, H. (1980). Genetical analysis of heading time by aid of close linkage with blast, *Pyricularia oryzae,* resistance in rice. *Bulletin of the National Institute of Agricultural Science* D31, 95-126.

Chapter 5

Botanic Gardens and Conservation

Robert O. Makinson

INTRODUCTION

There are approximately 2,200 botanic gardens in 153 countries, with an associated 142 million herbarium specimens and over 6 million living collection accessions (Wyse Jackson, 2001). About 70 percent of botanic gardens are in developed countries, and live plant and seed holdings tend to be strongly biased toward cool- and warm-temperate taxa. There are, however, appreciable and growing numbers of botanic gardens in developing countries with increasing capabilities to acquire and grow local species.

The collections, research programs, and expertise of botanic gardens can be an invaluable component of conservation genetics research and practice. Botanic garden collections vary widely in type (propagation material such as live plants and stored seed and identification resources such as herbarium specimens, DNA, and other phytochemicals) and quality (plants disease free, seed stored to maintain maximum viability). They also vary widely in their taxonomic and genetic representivity and degree of documentation (from none through variable reliability to extensive and fully reliable).

What Is a Botanic Garden?

The characteristics of botanic gardens are not straightforward. Botanic gardens are sometimes defined as "scientifically based" horticultural collections, although this may mean only that an attempt is made to correctly identify the plants and perhaps label them with scientific names for record keeping and public education purposes. It need not necessarily mean that identifications are correct, that there

doi:10.1300/5546_05

are authenticatable permanent voucher specimens, that there is any scientific research being conducted on the collections, or that documentation is sufficient for really sound research on them.

A related definition specifies botanic gardens as "comprehensive" or "representative" collections of living plants—perhaps across the plant kingdom where space and funds allow or perhaps themed to certain taxonomic groups or habitats or to plants related to certain human uses (e.g., medicinal or "physic" gardens). These two definitions, taken together, perhaps embody the most common perception of botanic gardens among horticulturalists and botanic scientists, and to some extent of the population at large.

The classic Kew model is that of a large metropolitan garden, often one established in the eighteenth or nineteenth century in part to exhibit the vegetable productions of the colonies and usually with a herbarium and a staff of taxonomic scientists dedicated to documenting the flora of those same colonies as part of the economic consolidation of colonial rule. As adjuncts to empire (often with a royal or presidential mandate), and because they were large tracts of open and usually public space in or near major cities, such gardens acquired and still retain considerable prestige.

The general public today, however, may have a different perception. Any horticultural space with a reasonably aesthetic appearance and selection of attractive plantings may qualify as a "botanic garden" in the eye of the public and of local authorities wishing to elevate their civic standing. Some countries have systems of accreditation, but these will usually have only partial relevance to the usefulness or otherwise of a garden for conservation purposes.

Any botanic garden, of whatever type, may have collections or expertise of significant conservation value. Many gardens have active research and conservation programs, and most have considerable circles of local or global botanic expertise available among their staff and external networks.

The Conservation Role of Botanic Gardens

A survey in the mid-1990s (summarized in Laliberte, 1997) found more than 1,700 botanic gardens and institutions worldwide holding plant collections that serve both conservation and educational purposes. Researchers should understand that botanic gardens almost always have multiple roles—any or all of the following: provision of public recreational amenity; delivery of primary, secondary, and tertiary education; research functions; horticultural development and training; events venue; land management; commercial business activities; and, perhaps explicitly, plant and sometimes animal conservation. This multiplicity of roles renders botanic gardens, large and small, vulnerable to periodic changes of priority and fluctuations of resources that may have a direct impact on the maintenance of the living plant material, herbaria, research programs, and information systems that are of most interest to the conservation researcher or practitioner.

In recent decades, as the scale of human-induced threat to natural biota has become more evident, many botanic gardens have adopted explicit statements of commitment to what is now termed "biodiversity conservation" as part of their fundamental mission. Some have played a leading role in promulgating the conservation objective to the global botanic gardens community. Most have at least a philosophical commitment to the idea of integrated conservation, i.e., that ex situ conservation is secondary to, and should serve the primary goal of, in situ conservation, including preservation of habitat. Genetics researchers who can demonstrate potential for in situ benefits from ex situ research are likely to secure more dedicated cooperation from both botanic gardens and permit authorities.

Key international agreements or documents in this process that are increasingly informing the specific programs of botanic gardens and sometimes of governments and that may provide a useful mandate for research projects include the following:

- *The International Agenda for Botanic Gardens in Conservation* (Wyse Jackson and Sutherland, 2000; http://www.bgci.org/ policies/index.html), launched at the World Botanic Gardens Congress in 2000
- *The Gran Canaria Declaration* (http://www.biodiv.org/doc/ meetings/cop/cop-05/information/cop-05-inf-32-en.pdf)
- *The International Plant Protection Convention* (IPPC) (http:// www.ippc.int/IPP/En/default.htm)
- *The International Treaty on Plant Genetic Resources of the Food and Agriculture Organisation* (http://www.bradford.ac.uk/ acad/sbtwc/gateway/TRADE/ITPGR.htm)

- The FAO *Global Plan of Action for the Conservation and Sustainable Utilisation of Plant Genetic Resources for Food and Agriculture* (http://www.fao.org/WAICENT/FaoInfo/Agricult/AGP/AGPS/GpaEN/gpatoc.htm)
- The *Plants Conservation Programme* of the IUCN Species Survival Commission (http://www.iucn.org/themes/ssc/plants/plantshome.html)
- *IUCN Technical Guidelines on the Management of ex-situ Populations for Conservation* (2000) (http://www.iucn.org/themes/ssc/pubs/policy/exsituen.htm)
- The *Global Strategy for Plant Conservation* (GSPC): Adopted as Decision VI/9 of the Conference of Parties (COP 6) to the Convention on Biological Diversity (CBD) in April 2002 at The Hague (http://www.biodiv.org/decisions/)

The GSPC is the latest of these and is intended to give a renewed impetus to the shaping and implementation of policies at the national level. Such policies are lacking or ineffective in many countries, particularly for wild plant conservation.

The GSPC is an agreement between governments adhering to the CBD and as such is part of a series of multilateral CBD-related commitments. The extent to which such international agreements will be effective, with direct influence on the policies of participating governments, has yet to be determined. The GSPC is part of a loosely meshing system of agreements and national/supranational legislation that include pre-CBD instruments such as CITES (Convention on International Trade in Endangered Species) and other CBD programs and plans related to biodiversity in the wild, sustainable use, and benefit sharing.

One of the GSPC objectives (Obj. 5[a] [iv]) is to

> Promote research on the genetic diversity, systematics, taxonomy, ecology and conservation biology of plants and plant communities, and the associated habitats and ecosystems, in situ (both in more natural and more managed environments), and, where necessary to complement in situ measures, ex situ, preferably in the country of origin. The Strategy will pay special attention to the conservation of the world's important areas of plant diversity, and to the conservation of plant species of direct importance to human societies.

The GSPC also set 16 targets for the period from adoption (in 2002) to 2010. Of particular relevance to conservation genetics are the following:

- *Target (2):* Perform a preliminary assessment of the conservation status of all known plant species at national, regional, and international levels;
- *Target (3):* Develop models with protocols for plant conservation and sustainable use based on research and practical experience;
- *Target (7):* Protect 60 percent of the world's threatened species in situ;
- *Target (8):* Maintain 60 percent of threatened plant species in accessible ex situ collections, preferably in the country of origin, with 10 percent of them included in recovery and restoration programs;
- *Target (9):* Conserve 70 percent of the genetic diversity of crops and other major socioeconomically valuable plant species conserved, and retain the knowledge of associated indigenous and local plants;
- *Target (15):* Increase the number of trained people working in appropriate plant conservation facilities according to national needs to achieve the targets of this strategy.

The implications and obligations of the GSPC are only beginning to influence signatory governments. It will be years before they are reflected in the practices and funding decisions of botanic gardens (many of which are partly or wholly dependent on nongovernment funding), even though botanic gardens were a critical part of the alliance of biodiversity organizations that negotiated the GSPC targets through to adoption.

It is important that conservation genetics research be identified where practicable within the framework of these and similar policies at national and subnational levels and that researchers work with botanic gardens and plant conservation organizations to publicize the linkage and the research results. Mutual reinforcement of this sort can have a positive influence on funding bodies and policy decision makers.

CRITICAL FACTORS IN USING BOTANIC GARDENS FOR CONSERVATION SCIENCE

The botanic garden resources of primary interest to conservation geneticists are their repositories of living biological material (whole live plants, stored seed, tissue banks, symbiont cultures), collections of preserved material for genetic analysis or for identification and validation of research (herbarium specimens of various sorts, tissue and DNA samples), the raw information associated with these collections (field labels, voucher history, ex situ management history), library collections, and human expertise (field knowledge of taxa, analytical knowledge of species relationships and ecology, and perhaps knowledge of past conservation actions).

Of these, the living and validation materials are likely to be the ones used in most studies. Their degree of usefulness for research and/or for recovery actions will depend both on biological and ecological factors relating to the taxon and on the ex situ history of the materials themselves and the reliability of their associated data.

Potentially, every ex situ collection of a wild taxon, or every exemplar of an "old variety" domesticated taxon, may be of conservation significance. If the only exemplars of a taxon are those retained in cultivation (the "extinct in the wild" conservation category), then even undocumented accessions are of high significance as long as they can be unambiguously identified. Conservation genetics may contribute to such identification or may depend upon it.

Botanic Gardens Data Systems

There are no universal or even very widespread formally adopted international standards for the capture of botanic gardens living collection data (comparable, for example, to the Herbarium Information Standards and Protocols for Interchange of Data, version 3, which is widely followed internationally). A degree of consistency in data concepts and data field definitions is nevertheless imposed simply by the similarities of work done by various botanic gardens, and over the past decade or so there has been a trend, particularly among better organized small- and medium-sized botanic gardens, to use one of a small number of "off the shelf" database systems customized for botanic gardens clients. Common use of such systems does not, however, imply a strict commonality of data field definitions, much less of primary data acquisition training and standards for collecting, accessing, and entering data.

Substantial work to establish an international standard has been done over the past 15 years. The International Transfer Format for Botanical Garden Records—Version 2.0 (ITF2) (TDWG, 1996) has been developed by the International Working Group on Taxonomic Databases for Plant Sciences (TDWG) of the International Union of Biological Sciences (IUBS).

Probably the major standardizing effect thus far of the ITF2 draft has been in the relatively widespread use of its data-dictionary by gardens using other home-grown or off-the-shelf databases. The latter include the proprietary product BG-BASE collections management software (http://rbg-web2.rbge.org.uk/bg-base), very popular with many botanic gardens records sections and compatible with ITF2 and several other TDWG standards.

Botanic gardens are likely to have modified such products, or to have developed their own databases, to suit local requirements and available labor resources for data capture and validation. The data fields neglected or deleted for these reasons are sometimes those that are highly relevant to the needs of conservation geneticists. Fields pertaining to identification of clonal lineages within living collections holdings are particularly subject to omission, neglect, or uneven attention. Training standards for data-entry staff also vary widely, and the incident of human error can be quite high. There are some guidelines documents regarding which data fields should be mandatory and which are highly reliable for some types of conservation actions, including those with a genetic component (BGCI, 1995; ANPC, 1997; Vallee et al., 2004).

Research Programs and Expertise

Many of the larger botanic gardens have active research programs, usually in the areas of taxonomy and evolutionary studies, ecology, plant breeding, phytogeography, and seed research. Any or all of these may be relevant to conservation geneticists, and there is great potential for productive formal and informal collaboration.

Gardens with research programs are usually open to and involved with tertiary student training. As specific knowledge of the biology, ecology, and/or taxonomy of organisms is usually required for sound conservation genetics research projects, gardens' co-supervision of students is often highly productive. Many larger gardens also have laboratory facilities, sometimes including DNA laboratories, that can serve as off-campus bases for students and visiting researchers. Gardens will usually be able to assist with advice on local requirements for collecting and research permits and may (subject to resources and permit conditions) be able to facilitate export of sample material. Visiting researchers should, however, be aware of the long lead times often necessary for acquisition of permits and of the resource constraints on many smaller gardens (sometimes including even for packaging materials and mail). It is the researcher's responsibility to negotiate arrangements well ahead of time, especially all permits, and budget for project support functions if necessary.

Nearly all gardens will have a considerable body of botanical expertise among their staff, advisory boards, and networks of local contacts and volunteers. These can be instrumental in assisting genetics researchers with accurate identifications, knowledge of significant populations of study taxa, provision of past field observations of ecology and demographics, and sometimes direct field support of specific projects.

Whole-Plant Live Collections

Usually termed *living collections* (seeds are living too, but are here referred to as *stored seed*), live whole-plant collections are usually the raison d'etre for a botanic garden. Each progeny line resulting from a specific collection event is typically referred to as an *accession,* and each accession will typically be assigned an *accession number* and have some form of *record* (manual or digital) recording its history.

There are, however, some important qualifications on the use of the term *accession.* An accession of a woody plant, acquired for propagation by transplant or cutting, will typically come from a single parent plant (a genetic individual), but this is not always the case—multiplant batches of propagation material may have been accessioned and maintained under one accession number. A common accession number should not be taken to indicate clonality. In contrast, accessions of herbaceous species are more frequently made up of material taken from multiple parent plants, of which some or all may have survived the propagation process and been maintained.

In either case, even when collection data explicitly state that collections under a single number are made from several parent plants, these may have been (unknown to the collector) clonal ramets. An evaluation of the probability of this (short of genetic analysis) can be made only by assessing the known biology of the species or population, from close examination of literature, herbarium specimens, and notes, from liaison with the collector (if possible), and from consultation with experts on the plant group concerned.

Most gardens records systems contain some errors resulting from vagaries of the initial collection data, errors in nursery handling and labeling during propagation and repropagation, data capture errors or omissions, or incorrect linkage to identification vouchers. Assumption of the uniclonality or otherwise of plants held under a single accession number in ex situ collections may thus present problems, and the plant should undergo genetic testing prior to use in major research or conservation management actions.

Original accessions obtained from seed rather than from vegetative material are usually reliably recorded as such. Plants resulting from ex situ repropagation by seed will generally be given a new accession number, as they are a new genetic individual. Because of the eccentricities of record system designs, however even plants resulting from vegetative repropagation events are sometimes given a new accession number. In either of these cases, there may or may not be a clear (and reliable) link between the original accession, intervening generations, the extant plant, and any permanent herbarium voucher or other samples. Close liaison with records officers is needed when herbarium and living collections systems are not closely coupled or when living collections documentation is obscure.

The acquisition, maintenance, genetic representivity, survivorship, health, and pedigree of ex situ plants are governed by multiple factors deriving from the numerous roles of botanic gardens, as noted previously. These factors impact upon the plants' usefulness for genetic studies. Usually, selection from wild populations will have been made on the basis of what constituted a "good horticultural selection," that is, on a particular factor likely to be maintained in cultivation, such as floral size, flower color, floriferousness, vegetative form, and robustness.

Only in a small minority of cases will living collections have been acquired and maintained with conservation research as a specific goal. More often, existing collections will become recognized by the botanic garden for its potential conservation significance, and this may or may not shape their subsequent maintenance. Usually, however, this recognition extends only to the level of threatened species—only rarely will intraspecific variation of any species be systematically a factor in the acquisition or maintenance of live plant collections, unless the garden concerned is closely involved with a research project or a species recovery plan.

The collections of interest to conservation geneticists are not necessarily limited to threatened species. Conservation actions for wild taxa increasingly focus, in the pursuit of best viability results, on ensuring appropriate provenance of material used for species restocking and translocation and for habitat regeneration and restoration programs. These programs are designed for common as well as endangered species or those that are only locally threatened. Broad-acre restoration actions often require either local-provenance propagation material (which may be hard to source in areas reduced to 2 to 3 percent native vegetation cover, as with much of the New South Wales South-west Slopes region, or the Western Australian wheat belt) or nonlocal-provenance material that is more tolerant of the habitat changes (accentuated summer and winter extreme temperatures, changed soil water conditions, soil salinity) that accompany the clearance of original vegetation cover. Effective selection and use of such material depends in part on sound genetic analysis. A final consideration is that research involving highly threatened species may first require study or experimentation on a common congeneric species as a surrogate before a more focused use of the scarce and valuable material is made.

Botanic Gardens Seed Banks

The issues faced by general seed banks are discussed in Chapter 4. Botanic garden seed banks are usually broad spectrum in terms of the taxonomic variety of their holdings. They are rarely equipped to accommodate truly representative infraspecific samplings of species, except for taxa of very restricted occurrence or of particular interest to that garden.

Nevertheless, the high volume of genetic variation in one seed batch (relative to the usually slim whole-plant holdings) makes any stock of seed potentially valuable. Laliberte (1997) notes that 30.7 percent of botanic gardens (from a survey of 1,500 institutions with 388 responses) have some form of cool- to low-temperature storage and over 250,000 germplasm accessions, and most have rare or threatened species.

As with whole-plant collections, caution needs to be exercised when evaluating seed bank accession records. Seed batch data will often but not always distinguish among seed collected from a single parent plant, that collected from multiple parent plants at the same site, and that collected from multiple plants at different sites (the last case will usually not be accessioned as a single batch under a single number). Handling errors may again confound even the best collection and data capture systems.

Viability and germinability testing histories for seed batches may or may not be available. For particularly valuable seed, attempts should be made to test germinability of seed of a close congeneric species that is of similar age and storage history before resorting to experimental work on the rare taxon.

Other Forms of Stored Living Material

Few botanic gardens maintain systematic in vitro tissue banks, although some material from species of active research interest may be kept. The vastly increased use of DNA in taxonomic work in the past decade, especially in many larger botanic gardens, means that quite large DNA banks, with intense sampling of species and genera, are sometimes held, and frozen source tissue may still be associated with these. The current rationale for maintaining these samples is for use in further taxonomic analyses or identification, but their potential as stock for genetic engineering and gene conservation is evident.

A surprising number of smaller gardens, often in developing countries, have active programs of maintaining field DNA banks or arboreta of selected species (living plants, usually separate from display collections), especially for long-lived perennials or taxa of economic interest and/or those with recalcitrant seeds or wholly vegetative reproductive modes. The Food and Agriculture Organization's *World Information and Early Warning System* (WIEWS) on plant genetic resources for food and agriculture is a useful resource (http://apps3. fao.org/wiews) for DNA banks and for stored seed information. Finally, some botanic gardens or their associated herbaria or research laboratories may maintain systematically sampled cultures of symbiont organisms (e.g., mycorrhizal fungi) for plant groups where these are significant, as with orchids.

WHO HAS WHAT SPECIES?

Because of the high level of awareness of conservation issues at most botanic gardens, known threatened species holdings are usually afforded some priority in the identification and data capture processes. Many gardens try to maintain at least some holdings of threatened species that occur within their area of taxonomic or geographical focus. Rarely, however, do they have the resources to acquire, or maintain indefinitely, collections that represent a substantive sample of the geographic range and genetic variation within a species.

Some botanic gardens have published hardcopy or Web lists of either their entire living collections holdings (by species) or of their holdings of rare and threatened taxa. Some attempts have also been made to compile such lists on a regional or national scale, as with the Web-based U.S. National Collection of Endangered Plants (http://www.centerforplantconservation.org/NC_Choice.html). Such information nodes need considerable resourcing to avoid becoming dated.

Seed lists are more commonly published than are whole-plant lists, and some may omit legislatively listed threatened species. Such lists may be useful as starting points for identifying holdings of interest. However, they must be supplemented because they are usually dated and the collections listed may not always have been critically identified against well-curated herbarium voucher specimens.

Botanic gardens will often have close links with dispersed collections of domesticated species (including old varieties of crop and horticultural taxa) held by enthusiast societies and individuals and are well-positioned to facilitate access. The resource constraints on most botanic gardens, and the usually unlinked nature of living collections and herbarium databases, however, mean that identifications in living collections tend to lag well behind current taxonomic nomenclature. The necessary sequence is one of critical screening of recent taxonomic literature, reidentification of herbarium vouchers and (ideally) of the actual living specimens, followed by updates of living collection databases and signage. Genetics researchers using ex situ material therefore should always allow time and resources within their projects to enable (1) checks (by themselves or by a taxonomic expert) of identifications of herbarium vouchers; (2) assessments of the reliability of the linkages between herbarium reidentification processes and living collections records (best achieved by frank discussions with records officers); and (3) a recheck by the researcher of a sample from the actual cultivated plant against reliably identified vouchers. Botanic gardens are invariably pleased to see their collections used for active research, and assistance in identifying misidentifications will be well received.

Direct liaison with living collections curators or records officers is mandatory to find suitable research material in ex situ collections. The taxon of interest will not necessarily be at the most geographically logical garden. The major botanic gardens and herbaria of any country will usually be able to refer genetics researchers to appropriate taxonomic experts and to some likely sources of living material beyond their own collections, but they may not have well-developed links with all smaller botanic gardens of possible relevance. It usually pays to seek more than one source of information.

Some excellent international and national level resources include the following:

- *Botanic Gardens Conservation International* (BGCI), Descanso House, 199 Kew Road, Richmond Surrey, TW9 3BW, United Kingdom; www.bgci.org
- *Centre for Plant Conservation,* Missouri Botanic Gardens, P.O. Box 299, St. Louis, MO 63166-0299, USA; phone: (314) 577-9450; e-mail: cpc@mobot.org; Web site: http://www.centerfor plantconservation.org
- *Directory of Australian Botanic Gardens and Arboreta* (J. Wilson & M. Fagg, eds.); Web site: http://www.anbg.gov.au/chabg/bg-dir/
- *Council of Heads of Australian Botanic Gardens,* Secretariat: CHABG, GPO Box 1777, Canberra ACT 2601, Australia; phone: +61 (0)2 6250 9507; e-mail: chabg-sec@anbg.gov.au; Web site: http://www.anbg.gov.au/chabg

- *Australian Network for Plant Conservation Inc.* (ANPC), GPO Box 1777, Canberra ACT, Australia; phone: +61 (0)2 6250 9509; e-mail: anpc@anbg.gov.au; Web site: www.anbg.gov.au/anpc

ACKNOWLEDGMENT AND PROMOTION OF EX SITU COLLECTIONS

Researchers utilizing ex situ collections should recognize the investment of time and labor involved to make them available. This needs to be acknowledged in resulting papers, but it is also helpful if the researcher takes the time to recontact the garden concerned after papers are finalized and discuss possible changes to the conservation significance and management of their collections, any unreliable data elements, and any "good stories" from the project that the botanic garden may be able to use toward media coverage or raising its conservation profile. Cultivation of long-term mutually beneficial relationships between researcher and garden is far more useful to the shared conservation effort than a "blow in, blow out" use of living collections.

Botanic gardens by nature have a high rate of public visitation and often a high level of political or civic interest in their activities. These factors open the door for both garden and researcher to showcase their involvement in research of direct conservation relevance.

PROPERTY RIGHTS, ACCESS, AND BENEFIT SHARING

The international botanic science community is heavily dependent on the free international flow of herbarium specimens, as loan and exchange, to facilitate basic taxonomic and diagnostic work. For this reason, herbaria and botanic gardens have been relatively proactive in developing strong self-regulatory controls over the movement of preserved specimens. They have also taken a close interest in the development of national and international regimes for the identification and protection of genetic and biological property rights and benefit sharing, the control through CITES of trade and exchange of endangered species, and in some countries and institutions in the issue of the intellectual and cultural property rights of indigenous peoples.

These are areas of rapidly developing national and international policy, with an overarching international legal framework embodied in the CBD and associated documents (http://www.biodiv.org). Assertion of national or subnational continuing property rights to biological material can potentially impede the development of public domain science, although most legal regimes allow continued exchange of material for noncommercial scientific or educational purposes. This does not remove the obligation on researchers and institutions to obtain prior informed consent for studies, declare any commercial interests or intentions, acknowledge countries of origin adequately, and maintain close control of such material (usually including an agreement not to pass living or derived material to third parties without consent of the country of origin).

The strict legal implications of the CBD relate only to material acquired since a 1993 starting date, which covers only about 10 percent of live plant holdings in botanic gardens (Wyse Jackson, 2001). However, many institutions are voluntarily adhering to the spirit of the CBD benefit-sharing provisions for material acquired before that date as well, and property rights legislation of various countries and subnational jurisdictions may be retrospective in the assertion of rights.

Researchers are well advised, when needing material of foreign origin, to seek creative ways of negotiating some form of benefit flow to the country of origin, even for nonremunerated and noncommercial research. This may take the form of scientific collaboration or training with country-of-origin personnel.

Increasing numbers of botanic gardens are adhering to the document *Principles on Access to Genetic Resources and Benefit-Sharing for Participating Institutions* (http://www.rbgkew.org.uk/conservation/principles.html), developed in the late 1990s to strengthen the self-regulatory regime of botanic gardens and herbaria in concordance with the CBD.

Researchers in conservation genetics may find that their ability to secure international access to living or voucher material is greatly enhanced by collaboration with a major botanic garden, as many of these have established routines of import/export permits, quarantine clearance, and the like. The obligation is on both garden and researcher to fully understand and comply with all relevant legal and policy instruments relating to genetic, biological, and intellectual property and the sharing of derived benefits. Facilitation by a garden does not mean that it can or will itself arrange for the necessary permits for a visiting researcher. It rarely has the legal power to do so, and researchers must be responsible themselves for securing all documents relevant to a project. Having a local botanic garden as a collaborating partner will generally help.

REFERENCES

ANPC (1997). "Guidelines for the translocation of threatened plants in Australia." Australian Network for Plant Conservation Inc., Canberra.

BGCI (1995). *A Handbook for Botanic Gardens on the Reintroduction of Plants to the Wild.* Botanic Gardens Conservation International, United Kingdom.

Laliberte, B. (1997). Botanic garden seed banks/genebanks worldwide, their facilities, collections and network. *Botanic Gardens Conservation News* 2 (9): 18-23.

TDWG (1996). International Working Group on Taxonomic Databases for Plant Sciences. *International Transfer Format for Botanical Garden Records—Version 2.0 [ITF2].* Unpublished. Available from Botanic Gardens Conservation International (BGCI), Descanso House, 199 Kew Road, Richmond Surrey, TW9 3BW, United Kingdom; www.bgci.org.

Vallee, L., Hogbin, T., Monks, L., Makinson, B., Matthes, M., and Rossetto, M. (2004). *Guidelines for the Translocation of Threatened Plants in Australia,* Second Edition. Australian Network for Plant Conservation, Canberra.

Wyse Jackson, P. (2001). An international review of the ex situ plant collections of the botanic gardens of the world. *Botanic Gardens Conservation News* 3 (6): 22-33.

Wyse Jackson, P.S. and Sutherland, L.A. (2000). *The International Agenda for Botanic Gardens in Conservation.* Botanic Gardens Conservation International, United Kingdom; http://www.bgci.org/policies/index.html.

Chapter 6

Conservation of Plant Genes and the Role of DNA Banks

Nicole Rice

INTRODUCTION

Australia is recognized as one of the most biologically diverse countries, with over 25,000 species of flowering plants. Australia, however, also holds the record for plant extinctions, with 7 percent of plant species considered to be at risk from human activities (Dixon, 2003). Extinction and loss of diversity are not problems limited only to Australia. Worldwide concern is now focused on the fate of natural resources and the potential for loss of biodiversity (Orr, 2003). Humans can control the fate of our genetic resources and the activities that threaten their sustainability. Activities and factors that pose threats include competition by introduced species, disease, pollution, agriculture, disturbance, and building and development (Guarino et al., 1995; Orr, 2003).

The conservation and sustainable use of plant genetic resources and biological diversity has been an international focal point and underpins the United Nations Convention on Biological Diversity (UNCED, 1992). The Convention on Biological Diversity also works to ensure that fair and equitable access is given to all plant genetic resources and that any benefits from use flow to the relevant parties. The Convention on Biological Diversity offers a compromise for conservationists and commercial sectors by encouraging controlled use of genetic resources and, when necessary, their protection. This, however, has not prevented the global loss of plant biodiversity, and it is evident that continued effort is needed to slow down the rate at which biodiversity is being lost (Dixon et al., 2003).

doi:10.1300/5546_06

DNA banks are discussed in this chapter as one approach to ex situ conservation of plant genes. Plant DNA banks are already established at both the Royal Botanic Garden Kew, United Kingdom (www. rbgkew.org.uk/data/dnabank/introcution.html) and the Missouri Botanic Gardens, United States (www.mobot.org/MOBOT/Research/ DNAdocs/dnabank1.html), and many private collections probably exist in association with plant molecular biology research programs. DNA banking is a relatively new technique for plant gene conservation, and operations at the Australian Plant DNA Bank are introduced and discussed.

CONSERVATION OF PLANT BIODIVERSITY

Plant conservation activities are expensive, and currently they are divided into in situ and ex situ approaches. In situ conservation strategies focus on the maintenance of plants in their natural environment, e.g., national parks and protected areas (Heywood, 1992; Given, 1994). In contrast, ex situ approaches conserve plants in an area foreign to their native habitat, e.g., in gardens or zoos (Heywood, 1992; Given, 1994). In situ and ex situ conservation activities are complementary, and it is essential that a variety of approaches be adopted to ensure that long-term conservation goals can be achieved (Bowen, 1999).

EX SITU COLLECTIONS

Ex situ collections try to maintain a wide range of the genetic diversity within each species (Hayward and Sackville Hamilton, 1997; Heywood and Iriondo, 2003). To represent the maximum amount of diversity and to document the collection in detail are priorities (Given, 1994). The techniques for ex situ conservation are discussed in more detail in Chapter 2.

The styles of ex situ collections in DNA banks are broadly classified into living and nonliving to reflect the potential end uses of the collection species (Table 6.1). Living collections maintain live plant material that can be used for future propagation of that species. Plant breeders use seed and germplasm collections to improve commercial crops. Herbarium and botanical voucher specimens are no longer

TABLE 6.1. Classification of Ex Situ Collections

Collection Type	Examples
Living collections *(material that is living or is stored to retain viability or to be propagated)*	Botanic gardens, arboreta
	Seed/germplasm collections
	Cyropreservation facilities
Nonliving collections *(material that is not living and is not stored to retain viability or to be propagated)*	Herbaria
	Extracted genomic DNA from plants

living and are rarely suitable for propagation even if seed has been collected. These specimens are maintained for the purposes of taxonomy, ecology, and documentation of the distribution of the species geographically (Crayn, 2003). Both living and nonliving collections store specimens that still contain DNA in their cells.

Germplasm collections consisting of seed and propagating materials are sometimes referred to as *gene banks*. Gene banks are collections of propagating materials that are checked for viability and stored under conditions that retain viability for long periods (Given, 1994). Miglani (1998, p. 106) also defines a gene bank as a "collection of world germplasm containing a large fraction of the total variability of a crop including its immediate relatives." The term *gene bank* can also be used to describe a collection of "extracted DNA" that was established as a result better molecular-based technologies (Mattick et al., 1992; Adams, 1997). DNA banks play an important role in the overall scheme of the conservation of species information and are key to completing an inventory of the total wealth of biodiversity. Collections of extracted DNA need to be defined separately to avoid confusion with existing germplasm collections.

HOW DO WE DEFINE A DNA BANK?

Both germplasm and extracted DNA collections preserve genes, and calling both collections *gene banks* leads to confusion. The main differences are how germplasm and DNA are stored and their end uses. To define a DNA bank, we need to consider the following terms:

Bank: A place for storing anything for future use (Hughes et al., 1995)

DNA: A polymer composed of deoxyribonucleotides linked together by phosphodiester bonds; the genetic material of the organism (Weaver and Hedrick, 1991)

Genomic DNA: One complete set of genetic information for the organism (Griffiths et al., 1999).

A plant DNA bank, therefore, conserves DNA that has been extracted from the plant cell. This DNA cannot, however, be used for propagation by conventional methods; it could be utilized only for plant propagation through biotechnology. It has been argued that future improvements in plant transformation techniques will enable ex situ collections to be used for genetic improvement, and thus the conservation of genetic diversity is essential (Godwin, 1997). In addition, DNA extracts should be conserved for the development of plant-based pharmaceuticals (Adams, 1997).

Two options for storage of DNA in a DNA bank have been proposed: (1) direct storage as extracts, libraries, or cloned fragments and (2) indirect storage as whole cells or frozen or desiccated tissues (Brown et al., 1997). Tissue, extracts, and gene complexes are ideal for prolonged storage of large numbers of samples securely, efficiently, and cheaply (Brown et al., 1997; Given, 1994). The dilemma facing ex situ collections is largely the cost of regeneration and monitoring of viability (Becker, 1998). The collection of wheat and corn at the International Maize and Wheat Improvement Center is estimated to cost $US8.86 million for 140,000 accessions, with an additional $US5.28 million if the free-of charge distribution is to be supported (Pardey et al., 2001). Although the proposal by Brown et al. (1997) to indirectly store DNA as whole cells or tissues would fulfill the requirements of a DNA bank, the DNA would need to be extracted upon request (Adams, 1997). A DNA bank run solely like this may be nothing more than a duplication of existing herbaria, botanic gardens, and in situ collections. Regardless of how the DNA is stored, a very strong commitment is required by the host institution to ensure adequate maintenance and sample growth of the collection (Adams, 1997).

An integrated program with both in situ and ex situ methods needs to be devised to ensure the long-term conservation of genetic resources (Given, 1994; Heywood and Iriondo, 2003; Heywood, 1992). A DNA bank would support one function of this holistic conservation model by providing a core collection of a wide range of DNA for molecular studies, including phylogenetics and systematics, biotechnology, and gene discovery (Adams, 1997; Bowen, 1999).

Operations at the Australian Plant DNA Bank

The Australian Plant DNA Bank (www.dnabank.com.au) was established in June 2001 and is hosted by the Centre for Plant Conservation Genetics, Southern Cross University, Northern New South Wales. The Australian Plant DNA Bank's main objective is to collect and store DNA from all species of Australian native flora for ex situ conservation. It will also store DNA of selected cultivated species. This is an ambitious task as there are approximately 25,000 indigenous species. It is also likely that multiple individuals from each species will be stored so as to capture the genetic diversity within and between various populations. The Australian Plant DNA Bank aims to store in excess of 25,000 unique accessions. It has also been suggested by Adams (1997) that collections should be opportunistic to avoid the extremely high cost of specific expeditions.

DATABASE AND SAMPLE MANAGEMENT

A major problem with ex situ/gene bank collections is a lack of documentation of samples (Given, 1994). The associated passport and characterization data are critical; without them, the samples have little or no value. Another problem is incorrect taxonomic identification of samples (Powledge, 1995; Given, 1994).

Data collection and their subsequent management are major concerns at the Australian Plant DNA Bank. It uses a relational database in Microsoft Access to store documentation for each sample (Exhibit 6.1). Samples are given a unique accession number, and the sample information is then posted on the World Wide Web (www.dnabank. com.au). Requestors can search the collection according to the scientific information, geographical area, and common name (Figure 6.1),

EXHIBIT 6.1. Information Collected for Samples in the Australian Plant DNA Bank

Core Sample Information

Accession number (identification number to track sample in database and collection)

Scientific/botanical information (family, genus, species, subspecies)

Variety and/or common name, photo, and Web links when available

Endangered status (according to Australian federal, state, and territory listings)

Restrictions on material (material transfer agreements, quarantine permits, etc.)

Donor Details

Donor name, address, etc. (person or group that deposited the sample)

Collection Details

Collection date, collector, site name and brief site description, longitude and latitude, region, state, and country

Population Details

Relative location of samples, height and diameter of specimen, flowers, etc. Population description (e.g., adults, juveniles,) other species, soil, and topography

Herbarium and Voucher Details

Herbarium or collection and voucher number

Reference Details

Author, journal, year, etc. (published reference related to the sample submitted)

Extraction Details

Material used (leaf, seed, etc.), method, DNA quantity and quality

FIGURE 6.1. Online Search Results from the Australian Plant DNA Bank Database, www.dnabank.com.au.

with access to the online database free of charge to registered users. Any extra information relating to a sample is held on file and stored by its accession number. When available, photographs of the specimens and links to other Web sites such as the herbarium or reference collection are included. Any native flora deposited must have been collected according to government legislation and the details concerning permission included.

One aspect of managing a collection and database is knowing any restrictions there may be on supplying DNA and/or associated information to requestors. There may be restrictions on the amount and type of information that is publicly available and on who is entitled to access. In the case of rare and endangered species, the exact locations of populations cannot by law be revealed. Even if the location details are manipulated to represent a region and not the exact geographical location, often a herbarium voucher will provide these details (Given, 1994). Communication of information at any level requires an enormous effort to ensure that informed consent is obtained and that the motives of the requestor are transparent. The Australian Plant DNA Bank stores a minimum of data with each sample and when possible links to other databases that have additional information. This has the added advantage of avoiding redundant information. For example, cultivated species sourced from Australian Genetic Resource collections in the Australian Plant DNA Bank have a link to the Australian Plant Genetic Resource Information Service (http://www.dpi.qld.gov.au/auspgris/), where additional information regarding the origin, collection details, and agronomic traits can be searched.

COLLECTION STRATEGIES

Collection of plant specimens is the core activity of any DNA bank. The need for adequate accompanying information has been emphasized, and the following procedures should govern the collection of samples for a DNA bank (Adams, 1997):

1. Taxonomically voucher all material in a herbarium.
2. Provide information on the locality and habitat of each sample.
3. Obtain correct permission from the country of origin, and adhere to all regulations.
4. Collect tissue and store in desiccant.
5. Provide photographs when possible.

For Australian native flora, the strategy for collection often varies and is primarily driven by the interests of the principal collecting group. Some species of native flora, e.g., *Angiopteris evecta* (the giant fern), have been deposited in the Australian Plant DNA Bank as part of a recovery planning process. Because there is only one remaining individual of this fern in NSW, the plant bank has captured the genetic diversity of the entire NSW population. For other rare species, there may be samples of every adult from the population(s), e.g., *Fontanea oraria*. In other cases, all populations in a geographical distribution may be extensively sampled but not every individual, e.g., *Davidsonia johnsonnii* (smooth leaf Davidson plum) and *Melaleuca alternifolia* (tea tree). Finally, some samples have been donated by research projects, e.g., *Elaeocarpus williamsiansus*.

The Australian Plant DNA Bank also tries to maintain the genetic diversity of cultivated species, but not to the same extent as for the native flora. Preference is given to sourcing of material from existing Australian and international genetic resource collections, and these collections then provide a reference similar to a herbarium voucher for native species. Wild relatives of crops are also included and are sourced from native Australian flora or again from genetic resource collections. For example, wild species of *Oryza* have been sourced from a germplasm collection (Tropical Crops and Forages Collection, Queensland Department of Primary Industries) in Queensland.

Material can be sourced directly from herbarium voucher specimens if necessary (Rogers and Bendich, 1985; Savolainen et al., 1995; Cubero et al., 1999; Taylor and Swann, 1994; Drábkova et al., 2002). However, this approach should be used with caution as the amount of material available for DNA extraction is obviously limited. One would also have to question the quality of the DNA obtained from herbarium specimens. Considerable effort has been invested to devise the best methods for preserving leaf material in the field for DNA extraction (Adams, 1997), and many voucher specimens may have been preserved before these methods were put into place. Plant material is often heated during the preservation process, and heat treatments above 55°C can make DNA extraction difficult (Adams, 1997).

Probably the most suitable way to preserve material for DNA extraction is cyropreservation (Adams, 1997); however, this method is not always cost-effective and is impractical for field collection. Samples collected in the field should be transported in silica gel (Adams, 1997, 1992 #197, [Chase, 1991]). Solutions of cetylmethylammonium bromide (CTAB) buffer (Štorchová et al., 2000) and alcohols are also effective transportation materials (Flournoy et al., 1996).

DNA EXTRACTION AND STORAGE

The Australian Plant DNA Bank generally uses one of two methods to extract DNA from plant tissue. The first method is a modification of the CTAB method of Maguire et al. (1994), which gives good results for Australian native and cultivated species. The second, although expensive, method is the commercial plant DNA extraction kit from QIAGEN, which yields good-quality DNA. Kit-based methods also often contain no organic solvents, making them easier to use.

For the Australian species, DNA is extracted from individual plant specimens and stored as a unique accession. This approach is not used for commercial crop species like wheat and barley. For these species, the leaf or seed material from a minimum of three individuals is pooled prior to extraction of the DNA.

The extracted genomic DNA are run on a 1 percent agarose gel to check the quality. Molecular weight size markers are run in conjunction with lambda DNA of varying concentrations. The absorbances at 260 and 280 nm are also recorded, and the ratio indicates the purity of the extracted. The extracts are then stored frozen at $-20°C$ and/or $-80°C$. Alternative methods for room temperature storage are currently under investigation (Adams, 1997). Finally, samples that are donated to the collection as DNA extracts are referenced to the published work of the donor and to his or her methodology.

CONSTRAINTS FOR EX SITU COLLECTIONS AND DNA BANKS

The constraints on DNA banks and other ex situ conservation activities are similar, e.g., high maintenance costs. All conservation activities need well-defined objectives, but often they are overambitious, not well planned, and poorly documented, often leading to their downfall (Adams, 1997; Given, 1994). DNA banks can be criticized for loss of higher level information, such as gene expression, development, and phenotypic attributes (morphology, secondary chemical compounds) (Brown et al., 1997). In addition, regeneration of DNA can be problematic. Regeneration of the whole genome by amplification is possible, but it may be biased. This is still under investigation (Brown et al., 1997).

Legislation and international agreements can also constrain operations of an ex situ collection. Since the Convention on Biological Diversity was implemented, access to genetic resources has become increasingly difficult. Some restrictions are necessary, however, to prevent overcollection of species in the wild, to protect the ecosystems in which they occur, and to ensure that overexploitation of resources is prevented and that financial returns from commercialization flow to the original owners. Many issues concerning ownership and access are still unresolved and will be under negotiation for some time. This has been confirmed by Kaplan (1998), who pointed out that it is now common for permission to collect plant species to be refused in some countries, because they believe they have not been compensated fairly in the past.

As a result of the international agreements, conservation of some plant genetic resources will largely lie with the country of origin. When such countries do not have the ability to conserve their own germplasm, they can collaborate with others to achieve this goal. On example is an arrangement between Paraguay and the United States, as outlined by Kaplan (1998).

CONCLUDING REMARKS

Conservation of plant genes is time-consuming and expensive. However, despite the ongoing problems, it is an important goal that requires a united effort and continued commitment. It is also critical that an integrated approach be adopted. DNA banks are effective repositories, and the extent to which they encourage research and use of wild species complements in situ conservation (Brown et al., 1997). The challenge now lies in the results—documenting diversity and ensuring that plant genetic resources are ultimately sustainable.

REFERENCES

Adams, R. P. (1997). Conservation of DNA: DNA banking. In *Biotechnology and Plant Genetic Resources: Conservation and Use,* Vol. 19 (Callow, J. A., Ford-Lloyd, B. V., and Newbury, H. J., eds.). CAB International, Oxon, UK, pp. 163-174.

Becker, H. (1998). Saving seeds for the long term. *Agricultural Research,* 46, 12-13.

Bowen, B. W. (1999). Preserving genes, species, or ecosystems? Healing the fractured foundations of conservation policy. *Molecular Ecology,* 8, S5-S10.

Brown, A. H. D., Brubaker, C. L., and Grace, J. P. (1997). Regeneration of germplasm samples: Wild versus cultivated plant species (Germplasm regeneration: Developments in population genetics and their implications). *Crop Science,* 37, 7-13.

Crayn, D. M. (2003). Vouchering: How and why. In *Plant Conservation Approaches and Techniques from an Australian Perspective* (Brown, C. L., Hall, F., and Mill, J., eds.). Australian Network for Plant Conservation, Canberra, Australia, module 3.

Cubero, O. F., Crespo, A., Fatehi, J., and Bridge, P. D. (1999). DNA extraction and PCR amplification method suitable for fresh, herbarium-stored, lichenized, and other fungi. *Plant Systematics and Evolution,* 216, 243-249.

Dixon, K. (2003). Australian Plant Biodiversity: A Framework for action. In *Plant Conservation Approaches and Techniques from an Australian Perspective* (Brown, C. L., Hall, F., and Mill, J., eds.). Australian Network for Plant Conservation, Canberra, Australia, module 1.

Dixon, K., Given, D., and Pearce, T. (2003). The global strategy for plant conservation. In *Plant Conservation Approaches and Techniques from an Australian Perspective* (Brown, C. L., Hall, F., and Mill, J., eds.). Australian Network for Plant
Conservation, Canberra, Australia, module 2.

Drábkova, L., Kirschner, J., and Vlcek, C. (2002). Comparison of seven DNA
extraction and amplification protocols in historical herbarium specimens of
Juncaceae. *Plant Molecular Biology Reporter,* 20, 161-175.

Flournoy, L. E., Adams, R. P., and Pandy, R. N. (1996). Interim and archival preservation of plant specimens in alcohols for DNA studies. *Biotechniques,* 20, 657-
660.

Given, D. R. (1994). *Principles and Practices of Plant Conservation.* Chapman and
Hall, London.

Godwin, I. D. (1997). Gene identification, isolation and transfer. In *Biotechnology
and Plant Genetic Resources: Conservation and Use,* Vol. 19 (Callow, J. A., Ford-
Lloyd, B. V., and Newbury, H. J., eds.). CAB International, UK, pp. 203-234.

Griffiths, A. J. F., Miller, J. H., Suzuki, D. T., Lewontin, R. C., and Gelbart, W. M.
(1999). *An Introduction to Genetic Analysis.* W. H. Freeman and Company, New
York.

Guarino, L., Rao, V. R., and Reid, R. (eds.) (1995). *Collecting Plant Genetic Diversity Technical Guidelines.* CAB International on behalf of the International Plant
Genetic Resources Institute (IPGRI), the Food and Agriculture Organization of
the United Nations (FAO), the World Conservation Union (ICUN), and the
United Nations Environment Programme (UNEP), Wallingford, UK.

Hayward, M. D. and Sackville Hamilton, N. R. (1997). Genetic diversity—Population structure and conservation. In *Biotechnology and Plant Genetic Resources:
Conservation and Use* (Callow, J. A., Ford-Lloyd, B. V., and Newbury, H. J.,
eds.). CAB International, Oxon UK, pp. 49-78.

Heywood, V. H. (1992). Efforts to conserve tropical plants—A global perspective.
In *Conservation of Plant Genes: DNA Banking and in Vitro Technology* (Adams,
R. P. and Adams, J. E., eds.). Academic Press Inc., San Diego, pp. 1-14.

Heywood, V. H. and Iriondo, J. M. (2003). Plant conservation: Old problems, new
perspectives. *Biological Conservation,* 113, 321-335.

Hughes, J. M., Michell, P. A., and Ramson, W. S. (eds.) (1995). *The Australian
Concise Oxford Dictionary.* Oxford University Press, Melbourne, Australia.

Kaplan, J. K. (1998). Conserving the world's plants. *Agricultural Research,* 46, 4-9.

Mattick, J. S., Ablett, E. M., and Edmonson, D. L. (1992). The gene library: Preservation and analysis of genetic diversity in Austalasia. In *Conservation of Plant
Genes: DNA Banking and in Vitro Biotechnology* (Adams, R. P. and Adams, J. E.,
eds.). Academic Press Inc., Royal Botanic Gardens, Kew, UK, pp. 15-36.

Miglani, G. S. (1998). *Dictionary of Plant Genetics and Molecular Biology.* The
Haworth Press, Inc., Binghamton, NY.

Orr, D. (2003). Diversity. *Conservation Biology,* 17, 948-951.

Pardey, P. G., Koo, B., Wright, B. D., Van Dusen, M. E., Skovmand, B., and Taba, S. (2001). Plant genetic resources. Costing the conservation of genetic resources: CIMMYT's ex-situ maize and wheat collection. *Crop Science,* 41, 1286-1299.

Powledge, F. (1995). The food supply's safety net. If a global agricultural crisis occurred, could the international germplasm community survive a run on its genebanks? *Bioscience,* 45, 235-243.

Rogers, S. O. and Bendich, A. J. (1985). Extraction of DNA from milligram amounts of fresh, herbarium and mummified plant tissues. *Plant Molecular Biology,* 5, 69-76.

Savolainen, V., Cuénoud, P., Spichiger, R., Martinez, M. D. P., Crèveocoeur, M., and Manen, J.-F. (1995). The use of herbarium specimens in DNA phylogenetics: Evaluation and improvement. *Plant Systematics and Evolution,* 197, 87-98.

Štorchová, H., Hrdlicková, R., Chertek, J., Tetera, M., Fitze, D., and Fehrer, J. (2000). An improved method of DNA isolation from plants collected in the field and conserved in saturated NaCl/CTAB solution. *TAXON,* 49, 79-84.

Taylor, J. W. and Swann, E. C. (1994). DNA from herbarium specimens. In *Ancient DNA: Recovery and Analysis of Genetic Material from Paleontogical, Archaeological, Museum, Medical and Forensic Specimens* (Herrmann, B. and Hummel, S., eds.). Springer, New York, pp. 166-181.

UNCED (1992). Convention on Biological Diversity. United Nations Conference on Environment and Development, Geneva.

Weaver, R. F. and Hedrick, P. W. (1991). *Basic Genetics.* Wm C Brown Publishers, Dubuque, IA.

Chapter 7

Strategies for in Situ Conservation

Ghillean T. Prance

INTRODUCTION

A dilemma that has faced conservationists for a long time is whether to focus on the conservation of species or of ecosystems. Initially much focus was given to endangered species and Red Data Books (Lucas and Synge, 1978). Recently the focus has definitely been more on ecosystems, and this has been strongly reinforced in the Convention on Biological Diversity (CBD) drafted in Rio de Janeiro in 1992. Ecosystem conservation is, of necessity, in situ, whereas species conservation can be carried out either in situ or ex situ depending upon the circumstances. Ex situ conservation is certainly useful and has saved a good number of species from extinction, but conservationists today favor the habitat approach whenever possible. The principal reasons why this is preferable are that the ecosystem approach conserves groups of interacting species, allows the process of evolution and adaptation to change, and often enables the protection of viable population sizes. Any strategy for in situ conservation will want to ensure that these processes continue. The ecosystem approach to conservation requires sensitivity to human activities in the area to be conserved. A balance between conservation and sustainable uses is required but not always easy to achieve.

Article 9 of the CBD states that ex situ conservation is predominantly for the purpose of complementing in situ measures. Article 8, with 13 clauses, outlines the rules for in situ conservation and, combined with Article 7 on identification and monitoring of areas, should form a basis for any strategy for in situ conservation. Some of the most important issues developed here are based on these requirements of

doi:10.1300/5546_07

the CBD, see the Appendix, which presents Article 8 of the CBD (UNEP/CBD, 1994).

IDENTIFICATION OF ECOSYSTEMS
TO BE CONSERVED

Crucial to the success of any strategy for in situ conservation is the selection of the appropriate areas to be conserved. Annex 1 of the CBD outlines the most important considerations:

1. Ecosystems and habitats: containing high diversity, large numbers of endemic or threatened species or wilderness; required by migratory species; of social, economic, cultural or scientific importance; or, which are representative, unique or associated with key evolutionary or other biological processes.
2. Species and communities which are: threatened; wild relatives of domesticated or cultivated species; of medicinal, agricultural or other economic value; or social, scientific or cultural importance; or importance for research into the conservation and sustainable use of biological diversity, such as indicator species.

This is a fairly comprehensive list, and the selection of areas to be conserved should be based on biological interest (diversity, endemism, and value) and on the level of threat to which an area is subject. The use of these types of indicators led to the identification of certain "hot spots" in the world that combined these categories (Myers, 1988; Myers et al., 2000), but these factors are equally important for local conservation strategies. The single most important category for selection of a conservation area is the relevance of the ecosystem to the regions, whether this be for biological reasons or for the preservation of resources to the community.

Research

The choice of priority areas for in situ conservation must be based on adequate scientific research that furnishes data on the characteristics mentioned previously. It is therefore necessary to carry out adequate inventory of the flora and fauna of the area or county under consideration. A survey of at least some key organisms is required to identify areas of species diversity and of endemism. Ethnobiological research is often needed to determine the social and cultural importance of an area or the presence of species of medicinal, agricultural, or other economic value. Research on the threats to an ecosystem are often as important as the gathering of the basic biological information. Overall the research must provide a strong and convincing logic for the conservation of an area.

Reserve Size

There has been much debate on the relative value of a single large reserve versus a number of smaller ones, the so-called SLOSS debate (Single Large or Several Small; Soulé, 1987). The answer to this will be based on the purpose of the conservation project. In many cases a single large reserve will be better because it preserves the large animals and is often less prone to extensive damage by fire, storm, or drought. However, sometimes a greater variety of habitats can be conserved through a network of smaller reserves.

Much research has been carried out on reserve size. For example, a project on the dynamics of forest fragments has been carried out over the past 25 years in an area of Amazonian rainforest just north of Manaus, Brazil (see Bierregaard et al., 1992; Laurance and Bierregaard, 1997). In this project reserves of different sizes (1, 10, 100, 1,000 ha, etc.) have been maintained in an area where the surrounding forest has been felled for agricultural purposes. The severe effect of isolation on small reserves is highly apparent from the study of many different organisms and of the physical environment. Severe damage to the edges of small reserves occurs through both susceptibility to storm damage and the arrival of invasive species. There is far less edge relative to areas in a large reserve, and so the "edge effect" (Lovejoy et al., 1986; Rankin de Merona et al., 1994) is of far less importance. Sometimes a network of smaller reserves can be made more effective through the maintenance of biological corridors between them. If there is free movement from one reserve to another of animals that pollinate or disperse seeds, then a small reserve can be efficient.

Population Size

A reserve needs to be large enough to conserve a viable population of the organisms it seeks to protect. There has been considerable debate about the minimum viable population (MVP) of a species (Soulé, 1987). For example, Laurence and Marshall (1997) suggested an MVP of 5,000 individuals, whereas Frankel and Soulé (1981) thought between 500 and 2,000 individual plants were necessary. Hawkes (1991) proposed an MVP of 1,000. Any strategy for in situ conservation will need to ensure that a viable population of the organisms to be conserved is included in the reserve. This will require different sizes of reserves for different organisms. Large animals at the top of the food chain will require much larger areas than smaller animals.

In the case of rainforest plants, the sizes of the populations of different species vary, and many species are quite sparsely distributed. Table 7.1 shows the data from an inventory of 4 ha of forest in western Amazonia. This area contains 3,158 trees of 10 cm or more in diameter that represent 556 species. Fourteen species with 25 individuals or more account for 786 individuals or 24.9 percent of the total. These species, especially the top three, *Eschweilera alba* (169 individuals), *Jessenia bataua* (107), and *Eschweilera odora* (99), would not need a large area to conserve a viable population of 5,000 individuals. At the bottom of the table are listed the 368 species with four, three, two, and one individual on the 4 ha. These rare species account for 66.2 percent of the total 556 species. A very large area would be needed to conserve a population of these species unless they occur elsewhere in large numbers. The increasing number of inventories in Amazonia (Ter Steege et al., 2003) show that this is unlikely. Therefore, for rainforest reserves, the larger the area can be, the more likely it will include viable populations of the trees.

Identification of Threats

It is most important to consider what threats are likely to adversely affect a conservation area. The CBD addressed this aspect in Article 7 c: "Identify processes and categories of activities which have, or are likely to have, significant adverse impacts on the conservation and sustainable use of biological diversity, and monitor their effects through sampling and other techniques." This aspect of reserve planning for in situ conservation of species is essential if a reserve is to be effective in the long term. The threats can come from many sources both human and biological, for example, industrial pollution of the air or water or the introduction of alien invasive species.

TABLE 7.1. Data from a Forest Inventory of 4 Hectares of Rainforest in the Rio Juruá Region of Amazonian Brazil

---

A. The 14 most abundant species

Species	No. of trees ≥ 10 cm diam
*Eschweilera alba** (= *E. grandiflora*)	169
Jessenia bataua (= *E. coriacea*)	107
*Eschweilera odora** (= *E. coriacea*)	99
Ragala sanguinolenta	54
Licania apetala	51
Micropholis guyanensis	49
Hevea pauciflora	44
Licania heteromorpha var. *heteromorpha*	40
*Eschweilera amara** (= *Lecythis idatimon*)	36
Pouteria guianensis	32
*Eschweilera fracta** (= *E. grandiflora*)	30
Guatteria poeppigiana	25
Iryanthera ulei	25
Saccoglottis guianensis	25
	786 of 3,158 trees

B. No. of individuals of the rarest species with 4 or less individuals/4 ha

1 indiv.   162
2 indiv.   99
3 indiv.   61
4 indiv.   46
Total   368 (66.2 percent of 556 species)

Source: Adapted from da Silva et al. (1992).

*Names of *Eschweilera* species corrected in accordance with Mori and Prance (1990).

In Hawaii, one of the greatest threats to the integrity of reserves is the constant arrival of alien plant species and of feral pigs and goats. The native flora of Hawaii comprise about 956 species, originating from between 270 and 280 original dispersals. However, the published flora of Hawaii comprise 1,678 species. The 722 species introduced by human activity have almost doubled the flora, and often drastic measures are needed to control or remove them from reserves. For example, in Kauai, alien species of ginger are being removed from the Kokee State Park through the use of chemical herbicides. Invasive species is one of the greatest threats to in situ conservation worldwide.

RELATIONSHIPS WITH THE LOCAL POPULATION

The conservation strategy for a reserve will work only when a good relationship has been established with the local population, whether indigenous or not. The CBD Article 8 i states, "Endeavour to provide the conditions needed for compatibility between present uses and the conservation of biological diversity and the sustainable use of its components." Article 8 j expands, ". . . respect, preserve and maintain knowledge, innovations and practices of indigenous and local communities embodying traditional lifestyles relevant for the conservation and sustainable use of biological diversity and promote their wider application with the approval and involvement of the holders of such knowledge, innovations and practices and encourage the equitable sharing of the benefits arising from the utilization of such knowledge, innovations and practices."

Without a close relationship with local people, neither conservation nor sustainable use can be achieved. The most successful reserves are ones where the local people are involved and benefit from the reserve. According to the concept of the biosphere reserve, developed by UNESCO (Batisse, 1986) to encourage this relationship, a core area allows access only to researchers and guards but the core is surrounded by a buffer zone in which economic activity can take place. Before any reserve is set up, the staff needs to establish a good strategy for the local people that addresses the socioeconomic aspects of the situation. This is also vital to the financing of a reserve, a topic that is discussed in a subsequent section.

The local political situation must also be respected. National, state, and municipal government officials as well as private individuals such as land owners may need to be involved. All interested parties should be catered to in a management strategy.

PROTECTION LEGISLATION

The need to establish adequate legislation is also addressed in the CBD. Article 8 b states, "Develop or maintain necessary legislation and/or other regulatory provisions for the protection of threatened species and populations." Without adequate legislative support, it can be impossible to control the use of a conservation area. Even where legislation exists, such as Brazil and Indonesia, it is still often impossible to preserve an area intact due to limited vigilance or to corrupt officials. However, once an area is legally established, it is easier to control and monitor and to exert international pressure for its maintenance.

MONITORING

An area protected for in situ conservation needs a strategy for monitoring. This is much more than just protection by guards. The status of the species being conserved needs ongoing monitoring to ensure that their populations are maintained and that biological interactions for pollination, seed dispersal, and other biological processes are maintained. In areas that are also used to support sustainable use, the harvestable species especially need careful monitoring. Species that are being harvested are at greater risk of population-level changes than are species in a pure biological reserve.

MAINTAINING THE ECOSYSTEM

In situ conservation focuses on maintaining a whole ecosystem rather than an individual species. Even when a reserve is established to focus on a particular species of animal or plant, an ecosystem approach is needed because the species of interest will be interacting with other species in its habitat. Any ecosystem is held together by a web of interaction between organisms for pollination, dispersal of seeds, defence from predators, and use of mycorrhizal fungi to absorb nutrients. Strategies for conservation need to consider all these aspects.

The Amazonian tree species *Couepia longipendula* (locally called *castanha da galinha*) produces a fruit from which a comestible oil is extracted by local people in Central Amazonia. The inflorescences hang down on long peduncles below the canopy (flagelliflory) and are visited by bats, which are its pollinators (Vogel, 1968/1969). The fruit then develop and later fall to the ground, where they are carried around and scatter-hoarded by agoutis. This economically useful tree species interacts with two mammals, one flying visitor to the flowers and the other a terrestrial disperser of the seeds. In addition, the roots of *C. longipendula* are infected with vesicular-arbuscular mycorrhiza that probably enhance the ability of this species to absorb nutrients. These types of interactions must be maintained for effective in situ conservation.

EFFECTIVE FINANCING

The more financially stable and independent a reserve is, the more likely it is to function long term. From the initial planning stage, it is important to have a sound strategy.

The extractivist reserves of Brazil were created by, and for, local people (Nepstad and Schwartzman, 1992; Wilkie, 1999). In these reserves the extraction of nontimber products such as rubber latex, Brazil nuts, and fruits is allowed, but the forest cannot be felled. These reserves are effectively slowing down deforestation in the states of Acre and Amapá, but their economies are marginal. The economies are much influenced by the market prices of any commodity that is extracted, and the value of rubber latex has fallen. These reserves have been useful, but their long-term viability is questionable. Where possible, the production of species of potential economic use needs to be encouraged.

Another common strategy for financing reserves is through ecotourism, and there are now many examples of where this is working. Ecotourism requires careful planning because it can also cause environmental damage. Nevertheless, tourism sustains conservation in many places, such as African game parks, the Galápagos Islands, and the Pantanal of Mato Grosso in Brazil. There are many smaller places where environmentally sensitive tourism has been established and also helps support reserves. In a small rainforest on the island of Savaii in Western Samoa, a rainforest canopy walk was built between two large trees in the Falealupo Rainforest Reserve. This is open to tourists for a fee, and their payment helps support this very small, but important reserve. The Falealupo Walkway was initially funded by a grant from a commercial company, Nu Skin International (see www. seacology.org; Cox, 1988; Cox and Elmqvist, 1997).

I have been involved in the establishment of the Reserva Ecologica Guapi-açu (REGUA) on the edge of the Organ Mountains in the state of Rio de Janeiro, Brazil. This reserve now protects over 5,000 ha of the much depleted Atlantic rainforest of Brazil. We were fortunate to be able to raise funds from donors to purchase a considerable area of land, to set up basic guarding by rangers, and to create education programs geared toward local schoolchildren. The purchased land also included a 450-ha farm. The farm buildings have been restored to make comfortable tourist facilities, and an area nearby has been restored to wetlands through the building of small dams. The wetlands are visible from the farm and enhance the visitor experience. In addition to the more luxurious farm buildings, another smaller accommodation has been built nearer to the forest for use by researchers and also tourists on lower budgets. The tourist side of REGUA is in the early stages of development, but from the start of planning one of the strategies to support this important reserve was through tourism. To date, 464 species of birds have been recorded at REGUA (www.regua.org).

A MANAGEMENT PLAN

All of the aspects of reserve planning mentioned must be brought together in a management plan for the reserve or conservation area (MacKinnon et al., 1986). Too many conservation areas proceed in a haphazard way. To be effective, a management plan must consider both the short-term and the long-term aspects. It needs to embrace all factors that could influence the future of the area and assign responsibilities to the various staff members. Other elements to be considered include methods of protection, monitoring, and research to be carried out and strategies to invovle the local population and to raise money. The latter should include a budget and designation of someone to oversee it. The purpose of a management plan is to guide the allocation of resources, set the priorities, and implement management actions.

CONCLUSION

In situ conservation is based on the ecosystem approach, which is a strategy for the integrated management of land, water, and living resources that promotes conservation and sustainable use of these resources in an equitable way (UNEP/CBD, 1994). This approach must address all aspects of the ecosystem processes and functions in order to succeed. It also must involve and sustain the local people. Its ultimate goal must be to preserve the ecosystem and the environmental services provided by that ecosystem.

APPENDIX:
ARTICLE 8 OF THE CONVENTION
ON BIOLOGICAL DIVERSITY (UNEP/CBD, 1994)

In Situ Conservation

Each Contracting Party shall, as far as possible and as appropriate:

(a) Establish a system of protected areas or areas where special measures need to be taken to conserve biological diversity;
(b) Develop, where necessary, guidelines for the selection, establishment and management of protected areas or areas where special measures need to be taken to conserve biological diversity;
(c) Regulate or manage biological resources important for the conservation of biological diversity whether within or outside protected areas, with a view to ensuring their conservation and sustainable use;
(d) Promote the protection of ecosystems, natural habitats and the maintenance of viable populations of species in natural surroundings;
(e) Promote environmentally sound and sustainable development in areas adjacent to protected areas with a view to furthering protection of these areas;
(f) Rehabilitate and restore degraded ecosystems and promote the recovery of threatened species, *inter alia,* through the development and implementation of plans or other management strategies;
(g) Establish or maintain means to regulate, manage or control the risks associated with the use and release of living modified organisms resulting from biotechnology which are likely to have adverse environmental impacts that could affect the conservation and sustainable use of biological diversity, taking also into account the risks to human health;

(h) Prevent the introduction of, control or eradicate those alien species which threaten ecosystems, habitats or species;

(i) Endeavour to provide the conditions needed for compatibility between present uses and the conservation of biological diversity and the sustainable use of its components;

(j) Subject to its national legislation, respect, preserve and maintain knowledge, innovations and practices of indigenous and local communities embodying traditional lifestyles relevant for the conservation and sustainable use of biological diversity and promote their wider application with the approval and involvement of the holders of such knowledge, innovations and practices and encourage the equitable sharing of the benefits arising from the utilization of such knowledge, innovations and practices;

(k) Develop or maintain necessary legislation and/or other regulatory provisions for the protection of threatened species and populations;

(l) Where a significant adverse effect on biological diversity has been determined pursuant to Article 7, regulate or manage the relevant processes and categories of activities; and

(m) Cooperate in providing financial and other support for in-situ conservation outlined in subparagraphs (a) to (l) above, particularly to developing countries.

REFERENCES

Batisse, M. (1986). Developing and focusing the biosphere reserve concept. *Nature and Resources* 22, 1-10.

Bierregaard Jr., R.O., Lovejoy, T.E., Kapos, V., dos Santos, A.A., and Hutchins, R.W. (1992). The biological dynamics of tropical rainforest fragments. *BioScience* 42, 859-866.

Cox, P.A. (1988). Samoan rainforest—partnership in the South Pacific. *National Parks* 62, 18-21.

Cox, P.A. and Elmqvist, T. (1997). Ecocolonialism and village controlled preserves in Samoa. *Ambio* 26, 84-89.

Frankel, O.H. and Soulé, M.E. (1981). *Conservation and evolution.* Cambridge University Press, Cambridge.

Hawkes, J.G. (1991). International workshop on dynamic in situ conservation of wild relatives of major cultivated plants: Summary of final discussion and recommendations. *Israel Journal of Botany* 40, 529-536.

Laurance, W.F. and Bierregaard Jr., R.O. (1997). *Tropical forest remnants: Ecology, management and conservation of fragmented communities.* University of Chicago Press, Chicago.

Laurence, M.J. and Marshall, D.F. (1997). Plant population genetics. In: Maxted, N., Ford-Lloyd, B.V., and Hawkes, J.G. (eds.), *Plant genetic conservation: The in situ approach.* Chapman Hall, London, pp. 99-113.

Lovejoy, T.E., Bierregard Jr., R.O., Rylands, A.B., Malcolm, J.R., Quintela, C.E., Harper, L.H., Brown Jr., K.S., Powell, A.H., Powell, G.U.N., Schubart, H.O.R., and Hays, M.B. (1986). Edge and other effects of isolation on Amazonian forest fragments. In: Soulé, M.E. (ed.), *Conservation biology: The science of scarcity and diversity.* Sinauer Associates, Sunderland, MA, pp. 257-285.

Lucas, G. and Synge, H. (1978). *The IUCN plant red data book.* IUCN, Morges, Switzerland, p. 540.

MacKinnon, J., MacKinnon, K., Child, G., and Thorsell, J. (1986). *Managing protected areas in the tropics.* IUCN/UNEP, Gland, Switzerland.

Mori, S.A. and Prance, G.T. (1990). Monograph of Lecythidaceae II. *Flora Neotropica* 21, 1-276.

Myers, N. (1988). Threatened biotas: "Hotspots" in tropical forests. *Environmentalist* 8, 187-208.

Myers, N., Mittermeier, R.A., Mittermeier, C.G., da Fonseca, G.A.B., and Kent, J. (2000). Biodiversity hotspots for conservation priorities. *Nature* 403, 853-858.

Nepstad, D. and Schwartzman, S. (1992). Introduction to non-timber product extraction from tropical forests: Evaluation of a conservation and development strategy. *Advances in Economic Botany* 9, vii-xii.

Rankin de Merona, J.M., Prance, G.T., Hutchins, R.W., da Silva, M.S., Rodrigues, W.A., and Vehling, M.E. (1994). Preliminary results of large-scale tree inventory of upland rainforest in Central Amazon. *Acta Amazonica* 22, 493-534.

Silva, A. S.L. da, Lisboa, P.L.B., and Maciel, U.N. (1992). Diversidade florística e estrutura em florresta densa da bacia do Rio Juruá - AM. *Bol. Mus. Paraense, Emílio Goeldi, Sér. Bot.* 8, 203-258.

Soulé, M.E. (ed.) (1987). *Viable populations for conservation.* Cambridge University Press, Cambridge and New York.

Ter Steege, H., Pitman, N., Sabatier, D., Castellanos, H., Van Der Hout, P., Daly, D.C., Silveira, M., Phillips, O., Vasquez, R., Van Andel, T., Duivenvoorden, J., Oliveira, A.A. de., Ek, R., Lilwhah, R., Thomas, R., Van Essen, J., Baider, C., Maas, P., Mori, S., Terborgh, T., Vargas, P.N., Mogollón, H., and Morawetz, W. (2003). A spatial model of tree α-diversity and tree density for the Amazon. *Biodiversity and Conservation* 12, 2255-2277.

UNEP/CBD. (1994). *The Convention on Biological Diversity: Text and Annexes.* Interim Secretariat for the Convention on Biological Diversity, Geneva, Switzerland.

Vogel, S. (1968/1969). Chiropterophilie in der Neotropischen. *Flora* 157, 562-602; 158, 195-202, 289-323.

Wilke, D. (1999). Carpe and non-wood forest products. In: Sunderland, T.C.H., Clark, L.E., and Vantomme, P. (eds.), *Non-wood forest products of Central Africa: Current research issues and prospects for development.* FAO, Rome, Italy, pp. 3-16.

Chapter 8

Impact of Habitat Fragmentation on Plant Populations

Maurizio Rossetto

INTRODUCTION

Natural ecosystems worldwide are subjected to increasing amounts of external pressure, degradation, and fragmentation with significant consequences to global biodiversity (Heywood, 1995). Yet, even natural landscapes are not always uniform, and many species live as metapopulations comprising an assemblage of discrete local subpopulations linked by dispersal and gene flow (Hanski and Gilpin, 1997). As a result, defining habitat fragments can be difficult, and researchers working on different organisms or ecosystems often rely on discrete interpretations (Watson, 2002). Generally, though, habitat fragmentation involves the artificial dissection of the natural distribution of a single species or of a group of species representing a functional ecological unit. Fragmentation can have considerable consequences on the population dynamics, genetic diversity, and overall fitness of species and poses serious threats to the integrity and long-term sustainability of the entire ecosystem. Because of the severity of its consequences, particularly for highly diverse ecosystems, habitat fragmentation is a main concern of the international conservation agenda.

Fragmentation into small, isolated remnants usually causes a decline in species richness and an increased vulnerability to external disturbances. This is particularly true at the interphase between the original ecosystem and the newly cleared area, where the "edge effect" is more prevalent. The term *edge effect* refers to the cumulative

doi:10.1300/5546_08

117

results of predation, competition, and abiotic decline that occur at the boundary between the ecosystem and the adjacent disturbed area (Saunders et al., 1991). As such boundaries are usually extensive in habitat fragments, the edge effect can have significant consequences on their biodiversity.

Biodiversity conservation within fragmented habitats depends on the capacity of single populations to survive bottlenecks and stochastic events. Understanding the changes in genetic diversity and gene flow that follow the dissection and isolation of populations is an important step toward the development of better management strategies. Accordingly, it is now widely recognized that realistic population viability analysis (PVA) models should include genetic and population dynamics information (Menges, 2000). However, as suggested by Young et al. (1996), many factors contribute to the final trajectory followed by fragmented populations, and, while it can be assumed that genetic changes will occur, the nature of these changes cannot always be predicted by simple theory. Fortunately, a range of highly informative molecular approaches are making the collection of critical genetic information on variability, inbreeding rates, and gene flow increasingly attainable.

This chapter briefly examines some of the major genetic concerns associated with habitat fragmentation and what consequences they can have on population viability. It is hoped that the growing pool of conservation genetic studies will ultimately lead to the development of broadly applicable models for the management and conservation of fragmented habitats.

CONSEQUENCES OF HABITAT FRAGMENTATION: GENETIC HAZARDS SMALL, ISOLATED POPULATIONS FACE

Landscape fragmentation initiates losses of habitat types, causes declines in population sizes, and increases isolation. A direct consequence of these changes is the reduction in genetic diversity and gene flow within and among remnants. Ultimately, a number of genetic processes will lead to inbreeding depression, eventually reducing fitness, adaptation potential, and viability of the metapopulation.

Loss of Evolutionary Potential

Habitat fragmentation divides large, continuous populations into small, isolated remnants with reduced diversity and little gene exchange. In the absence of gene flow, within each reproductive cycle a portion of the alleles present within the parental individuals may, purely by chance, not be passed on to the following generation. Genetic drift, the cumulative effect of these frequency changes, will eventually erode diversity through allelic fixation. In the absence of gene flow, mutation would be the main mechanism capable of restoring diversity, but in small populations mutations occur too infrequenly to be beneficial. Consequently, the overall loss of allelic diversity is often a typical outcome of population fragmentation.

As environmental degradation causes a founder effect within each fragment, rare alleles are usually lost immediately. Further allelic diversity can then be lost through random fluctuations in allelic frequencies from one generation to the next. Significant reductions of allelic richness in fragmented habitats have been observed in a range of species, including forbes, *Salvia pratensis* and *Scabiosa columbaria* (Van Treuren et al., 1991) and *Rutidosis leptorrhynchoides* (Young et al., 1999), open woodland trees, *Eucalyptus albens* (Prober and Brown, 1994), and rainforest trees, *Swietenia humilis* (White et al., 1999). In the long term, such losses can diminish the ability of populations to evolve in response to changes in environmental conditions and reduce the overall gene pool available to the species.

Quantifying the short-term evolutionary outcomes caused by the decline in allelic diversity that follows habitat dissection can be difficult. Nonetheless, there are instances in which loss of allelic diversity can have an immediate impact on the reproductive potential of single populations. Young et al. (2000) found that for the self-incompatible *Rutidosis leptorrhynchoides,* an endangered daisy, reductions in population size corresponds to a decline in S allele richness. As the number of S alleles decreases, so does the effective population size because fewer compatible mates become available. This can trigger a reduction in reproductive output and can also increase the success of those individuals with the less common S alleles. Habitat fragmentation can also cause immediate demographic shifts in outcrossing species capable of vegetative reproduction. A molecular study on an endangered rainforest tree, *Elaeocarpus williamsianus,* found that habitat clearing destroyed most of the genetic diversity within the remaining populations, with single clones found at six out of seven sites (Rossetto, Gross, et al., 2004). As *E. williamsianus* is a preferential outcrosser, viable seed are only found within the multiclonal population. Sexual reproduction is highly unlikely at the other sites, a predicament that significantly hinders the potential of the isolated clones for recombination and long-distance dispersal.

Changes in allelic frequencies are often detected among recently fragmented populations, but changes in genetic variation (measured as expected heterozygosity, H_e) are more difficult to identify. Populations of *Swietenia humilis* that were dissected approximately 50 years ago (a short time for a long-lived rainforest tree) showed significant losses of rare alleles but not of genetic variation (White et al., 1999). This is because changes in heterozygosity levels are directly dependent on the effective population size during the bottleneck: $\Delta H = H_1 - H_0 = -(1/2N_e) H_0$, where H_0 and H_1 are the levels of heterozygosity before and after the bottleneck, respectively. Thus, a bottleneck of $N_e = 50$ reduces heterozygosity by only 1 percent. This implies that significant losses in genetic variation occur only when N_e is consistently low across successive generations, or when there is an extreme bottleneck. It is important to note, though, that because the loss of diversity is directly related to the smallest bottleneck N_e, even if a population quickly recovers after being fragmented and isolated, its genetic variation will be directly influenced by the initial number of its founders.

Increased Inbreeding Rates

Inbreeding, the change from random mating to mating between increasingly related individuals, is identifiable by lower levels of observed than expected heterozygosity. In small, isolated populations inbreeding is unavoidable as all individuals will eventually become related by descent. The ensuing loss of heterozygosity can have immediate consequences to the survival potential of these populations. A decline in reproductive efficiency, seed germination failure, and reduced offspring survival rates are some of the possible consequences of inbreeding and are cumulatively known as *inbreeding depression.*

Inbreeding depression can be caused by the increased frequency of recessive and partially recessive alleles or by the breakdown of overdominance mechanisms (which arise where heterozygous genotypes provide adaptational advantages). This can reduce individual fitness, which is measured by the quality and quantity of its reproductive contribution. Various approaches exist for expressing the extent of inbreeding depression based on quantitative data and lethal equivalent measures (Lynch and Walsh, 1998). Pollen quality, ovule and seed numbers, germination rates, and competitive ability are among the viability characteristics that can be negatively affected by inbreeding (Frankham et al., 2002). For example, *Silene alba* crosses representing a number of widely separated locations showed a negative correlation between inbreeding coefficient and seed germination rate (Richards, 2000). In most cases loss of fitness is caused by the additive effects of a large number of loci and can be further accentuated by stochastic environmental changes brought about by habitat fragmentation. Thus, ideally inbreeding depression should be measured from natural populations within local environmental conditions, but in situ studies are often difficult to carry out and are still uncommon.

Some plants are less affected than others by the consequences of inbreeding. A negative correlation between inbreeding depression and selfing rates was reported by Husband and Schemske (1996). This is presumably due to the fact that selfers have had more opportunities to select against deleterious recessives. Polyploidy can also mitigate the intensity of inbreeding depression, as the presence of multiple alleles can mask the expression of deleterious recessives and reduce drift rates. Occasionally inbreeding does not reduce fitness but facilitates the removal (purging) of deleterious recessive alleles by natural selection. Purging can be highly efficient with recessive lethal or detrimental alleles with large effect, but less so with alleles with a smaller effect. As a result, the circumstances leading to the effective removal of the genetic load have been considered as unlikely in most field populations (Keller and Waller, 2002).

Because mutation is unlikely to increase diversity in small, fragmented populations, the only process capable of reversing the effects of inbreeding depression is outbreeding. Outbreeding involves the exchange of genes between fragments (or with undisturbed populations) and is particularly effective when a heterogenous group of remnants is created by random genetic drift. A number of mechanisms through which "island"/"sink" populations can exchange gene diversity with each other or with "mainland"/"source" populations have been proposed (Hedrick, 1983), but ultimately what matters to the viability of small fragments is the level of external gene flow and the factors restricting it.

The changes in inbreeding levels, drift, and gene flow caused by habitat fragmentation can be estimated using Wright's F-statistics (1969) or one the many equivalent analytical approaches. Inbreeding estimates depend on a reference population and can be summarized by $(1 - F_{it}) = (1 - F_{is})(1 - F_{st})$ (Keller and Waller, 2002). F_{it}, the total inbreeding value, represents pedigree inbreeding averaged over all individuals and measures the amount of ancestry that is shared by their parents. Within isolated remnant populations, F_{it} levels are usually high due to increased biparental inbreeding. F_{is}, the inbreeding coefficient, represents inbreeding measured as nonrandom mating and refers to the increased degree of relatedness between two mates relative to two mates chosen at random from the population. As just mentioned, in fragmented populations even random matings can result in matings between relatives; thus F_{is} can be zero despite high F_{it} levels. Finally, F_{st}, the fixation index, measures the levels of inbreeding caused by population subdivision. As gene flow is reduced among isolated fragments, reduced N_e and genetic drift will result in population differentiation and higher F_{st} values.

Alterations in Population Dynamics and Gene Flow

The reduction in gene flow that follows habitat fragmentation can cause losses of diversity, drift, inbreeding, and increased genetic differentiation among remnants. Therefore, in order to predict the long-term viability of the fragmented metapopulation, it is important to estimate the quantity and quality of genetic exchange that is taking place. Migration rates between populations (N_m) can be estimated as number of migrants per generation using measured F_{st} values in the following equation: $N_m = [(1/F_{st}) - 1] / 4$. In theory, one migrant per generation is sufficient to prevent population substructure (Hartl and Clark, 1997), but a greater number is likely to be needed for sufficient between-population exchange to be ensured (Frankham et al., 2002).

An ideal approach for investigating the consequences of habitat fragmentation is to compare and contrast levels of gene exchange in fragmented and continuous populations. It is important to consider the effect that overlapping generations can have on the measures of gene flow obtained from fragmented populations. Increased differentiation among fragments can be concealed by the presence of predisturbance individuals, particularly in long-lived species such as tropical trees (White et al., 1999). However, even if the persistence of preclearing individuals influences the overall results, it can be expected that younger cohorts will display greater interpopulation differentiation than older plants. For instance, a demographic study on *Symphonia globulifera* detected greater differentiation among juvenile plants within fragmented rainforest in Costa Rica (Aldrich et al., 1998). Overlapping generations can also have a significant effect on shorter-lived species, particularly those relying on long-lived seed banks. As little differentiation was detected among small, fragmented populations of *Grevillea scapigera*, a short-lived heathland shrub with a long-lived seed bank, it was concluded that the measured diversity and lack of population structure reflected preclearing gene flow still characterized within the original seed bank (Rossetto et al., 1995).

Natural historical fragmentation events can also influence the levels of gene diversity and gene flow measured across populations. For instance, bottlenecks during the last glacial period were suggested as the main cause for the differentiation measured among fragmented populations of *Grevillea macleayana* (England et al., 2002). Similarly, a recent study on *Elaeocarpus grandis* detected analogous levels of genetic diversity among fragmented and undisturbed population and no changes in population structure across age cohorts (Rossetto, Jones, and Hunter, 2004). It was concluded that the effect of severe population bottlenecks during the last glacial period concealed some of the changes expected from more recent rainforest clearing events.

Because of the numerous external and historical factors that can influence the overall levels of inbreeding within populations, it is often more informative to use direct rather than inferred methods of measuring gene flow. The advantage in using direct gene flow measures is that they provide data on current rather than equilibrium level gene exchange. Direct measures of gene flow require the genotyping of all reproductive adults within a selected area, for example, a small remnant. A progeny array from a selected number of individuals is then genotyped, and maximum likelihood methods are used to assign parentage to each seed or seedling. If parentage cannot be assigned locally, the genotypic information can be used to determine the interpopulation origin of the progeny. If external gene flow has been identified, the comparative use of maker systems based on nuclear or plastid DNA can help distinguish between pollen or seed dispersal (Wang and Smith, 2002). This approach will provide relevant data on the average distance travelled by pollen within and between sites and on the contributions of the parental cohort to the following generations.

Direct gene flow measures can provide new information on how fragmentation influences the movements of pollinators and dispersal organisms and, consequently, on the potential for recovery of single remnants. Traditional on-site ecological approaches for determining dispersal and migration can be extremely difficult and often lead to underestimates. With the advent of highly informative molecular techniques such as microsatellites, tracking interpopulation movement based on genotypic information alone has become increasingly popular (Ouborg et al., 1999). Direct gene flow measures are particularly valuable when comparing recently fragmented to continuous populations. Current studies have shown that pollen can travel greater distances than previously expected, implying that small fragments or even isolated plants can have an important role within the metapopulation. Nason et al. (1998) found that the small wasps responsible for the pollination of strangler figs can travel distances in excess of 14 km. A recent study on *Dinizia excelsa* showed how the combination of habitat disturbance and the introduction of a new pollinator to the region, the African honeybee, expanded genetic neighborhoods within this close-canopy tropical tree (Dick et al., 2003). Greater-than-expected pollen exchange has also been measured for wind-pollinated species such as *Quercus macrocarpa* and *Pinus densiflora* (Dow and Ashley, 1996; Lian et al., 2001).

Parentage analysis studies can be extremely informative, but they also require wide-ranging sampling of progenies and of putative fathers and consequently are labor and resources intensive. Fortunately, new analytical procedures that do not require the high level of resolution needed by parentage analysis are regularly being developed. TwoGener is a two-generation (parent-offspring) approach allowing the quantification of heterogeneity among male gamete pools sampled by maternal plants scattered across the landscape (Smouse et al., 2001). TwoGener enables the estimation of mean pollination distance and effective neighborhood size and, despite sacrificing paternal designation, provides an understanding of gene flow at a landscape scale with a moderate investment in resources. Using this approach, Sork et al. (2002) detected a decline in the number of fathers contributing to the progeny of single maternal *Quercus lobata* trees within recently fragmented Californian populations. Because estimators such F_{st} may not reflect current gene exchange patterns of gene flow within recently fragmented habitats, autocorrelation approaches (such as the microspatial autocorrelation analysis developed by Smouse and Peakall, 1999) can also provide greater detail on the relationships among spatially structured individuals and on localized inbreeding events.

CONCLUSION: DEVELOPING INFORMATIVE HABITAT FRAGMENTATION RESEARCH

Highly informative molecular techniques and increasingly sophisticated analytical approaches have given new impetus to habitat fragmentation research. With better data gathering techniques, it is now possible to learn more about the likely long-term changes caused by environmental degradation. By selecting appropriate research matrices and representative study species, it will eventually be possible to develop broad conservation and management strategies. Following is a brief list of issues worthy of consideration when planning research on fragmented habitats.

Study System:

- Prioritize habitats containing high levels of biodiversity and/or are restricted to less than 10 percent of their initial distribution.
- Sample a sufficient number of sites to represent various grades of fragmentation (such as different population size and isolation classes) and compare to control populations within continuous habitats.
- If available, include selected experimental systems such as areas linked by corridors or ones that are artificially revegetated.
- Select representative species occupying important ecological niches within the ecosystem of interest. A multispecies approach providing directly comparable information across a number of taxa is preferable.

- When possible, select taxa with known breeding system and dispersal vectors (for both pollen and seeds). This will enable a comparison between contrasting systems and some generalization based on specific biological attributes.

Sampling Approach:

- Collect ecological and demographic information across all selected sites.
- Distinguish age cohorts among sampled individuals for analytical differentiation between postfragmentation juveniles and prefragmentation adults. Where seed banks exist, compare these and current seed crops.
- Sample a sufficient number of individuals to enable direct gene flow studies, at least within representative sites.
- Ideally, repeated sampling of juveniles over multiple seasons will provide more accurate information, because they are less likely to be biased by periodic environmental events.
- Consider the existence of potentially different management units when selecting sampling sites.

Analytical Approach:

- Use highly informative molecular markers such as microsatellites, and, if possible, compare data from nuclear and plastid DNA. Gathering information from more than one class of molecular markers will further clarify the biological history of the metapopulation.
- Use both indirect and direct methods to measure changes in pollen and seed dispersal rates that have occurred as a result of environmental degradation.
- Obtain comparative measures of inbreeding depression in situ or experimentally. Measures of reproductive success can be obtained directly from the populations studied, although local environmental conditions can influence the outcomes. Alternatively, the fitness of inbred versus outbred individuals can be tested in glasshouses using controlled environmental conditions.

These suggestions are not always applicable or practical due to specific research constraints, but they represent some of the main directions of current habitat fragmentation studies. As habitat fragmentation and environmental degradation become more prevalent, conservation and land management agencies are increasingly concerned with the conservation of fragmented rather than pristine habitats. The availability of directly comparable data on the overall influence that such changes have on the environment will ultimately lead to broadly applicable conservation and management strategies.

REFERENCES

Aldrich, P.R., Hamrick, J.L., Chavarriaga, P., and Kochert, G. (1998). Microsatellite analysis of demographic genetic structure in fragmented populations of the tropical tree *Symphonia globulifera. Molecular Ecology* 7, 933-934.

Dick, C.W., Etchelecu, G., and Austerlitz, F. (2003). Pollen dispersal of tropical trees (*Dinizia excelsa*: Fabaceae) by native insects and African honeybees in pristine and fragmented Amazonian rainforest. *Molecular Ecology* 12, 753-764.

Dow, B.D. and Ashley, M.V. (1996). Microsatellite analysis of seed dispersal and parentage of saplings in bur oak, *Quercus macrocarpa. Molecular Ecology* 5, 615-627.

England, P.R., Usher, A.V., Whelan, R.J., and Ayre, D.J. (2002). Microsatellite diversity and genetic structure of fragmented populations of the rare, fire-dependent shrub *Grevillea macleayana. Molecular Ecology* 11, 967-977.

Frankham, R., Ballou, J.D., and Briscoe, D.A. (2002). *Introduction to conservation genetics.* Cambridge: Cambridge University Press.

Hanski, I.A. and Gilpin, M.E. (1997). *Metapopulation biology: Ecology, genetics and evolution.* London: Academic Press.

Hartl, D.L. and Clark, A.G. (1997). *Principles of population genetics.* Sunderland, MA: Sinauer Associates.

Hedrick, P.W. (1983). *Genetics of populations.* Boston: Science Books International.

Heywood, V.H. (1995). *Global biodiversity assessment.* Cambridge: Cambridge University Press.

Husband, B.C. and Schemske, D.W. (1996). Evolution of the magnitude and timing of inbreeding depression in plants. *Evolution* 50, 54-70.

Keller, L.F. and Waller, M. (2002). Inbreeding effects in wild populations. *Trends in Ecology and Evolution* 17, 230-241.

Lian, C., Miwa, M., and Hogetsu, T. (2001). Outcrossing and paternity analysis of *Pinus densiflora* (Japanese red pine) by microsatellite polymorphism. *Heredity* 87, 88-98.

Lynch, M. and Walsh, B. (1998). *Genetics and analysis of quantitative traits.* Sunderland, MA: Sinauer Associates.

Menges, E.S. (2000). Population viability analyses implants: Challenges and opportunities. *Trends in Ecology and Evolution* 15, 51-56.

Nason, J.D., Herre, E.A., and Hamrick, J.L. (1998). The breeding structure of a tropical keystone plant resource. *Nature* 391, 685-687.

Ouborg, N.J., Piquot, Y., and Van Groenendael, J.M. (1999). Population genetics, molecular markers and the study of dispersal of plants. *Journal of Ecology* 87, 551-568.

Prober, S.M. and Brown, A.D.H. (1994). Conservation of the grassy white box woodlands: Populations genetics and fragmentation of *Eucalyptus albens*. *Conservation Biology* 8, 1003-1013.

Richards, C.M. (2000). Genetic and demographic influences on population persistence: Gene flow and genetic rescue in *Silene alba*. In *Genetics, demography and viability of fragmented populations* (pp. 271-291), Young, A.G. and Clarke G.M. (eds.). Cambridge: Cambridge University Press.

Rossetto, M., Gross, C.L., Jones, R., and Hunter, J. (2004). The impact of clonality on an endangered tree *(Elaeocarpus williamsianus)* in a fragmented rainforest. *Biological Conservation* 117, 33-39.

Rossetto, M., Jones, R., and Hunter, J. (2004). Genetic effects of rainforest fragmentation in an early successional tree *(Elaeocarpus grandis)*. *Heredity* 93, 610-619.

Rossetto, M., Weaver, P.K., and Dixon, K.W. (1995). Use of RAPD analysis in devising conservation strategies for the rare and endangered *Grevillea scapigera* (Proteaceae). *Molecular Ecology* 4, 321-329.

Saunders, D.A., Hobbs, R.J., and Margules, C.R. (1991). Biological consequences of ecosystem fragmentation: A review. *Conservation Biology* 5, 18-32.

Smouse, P.E., Dyer, R.J., Westfall, R.D., and Sork, V.L. (2001). Two-generation analysis of pollen flow across a landscape: I. Male gamete heterogeneity among females. *Evolution* 55, 260-271.

Smouse, P.E. and Peakall, R. (1999). Spatial autocorrelation analysis of individual multiallele and multilocus genetic structure. *Heredity* 82, 561-573.

Sork, V.L., Davis, F.W., Smouse, P.E., Apsit, V.J., Dyer, R.J., Fernandez, J.F., and Kuhnm, B. (2002). Pollen movement in declining populations of California Valley oak, *Quercus lobata:* Where have all the fathers gone? *Molecular Ecology* 11, 1657-1668.

Van Treuren, R., Bijlsma, R., Ouborg, N.J., and van Delden, W. (1991). The significance of genetic erosion in the process of extinction: I. Genetic differentiation in *Salvia pratensis* and *Scabiosa columbaria* in relation to population size. *Heredity* 66, 181-189.

Wang, B.C. and Smith, T.B. (2002). Closing the seed dispersal loop. *Trends in Ecology and Evolution* 17, 379-385.

Watson, D.M. (2002). A conceptual framework for studying species composition in fragments, islands and other patchy ecosystems. *Journal of Biogeography* 29, 823-834.

White, G.M., Boshier, D.H., and Powell, W. (1999). Genetic variation within a fragmented population of *Swietenia humilis* Zucc. *Molecular Ecology* 8, 1899-1909.

Wright, S. (1969). *Evolution and the genetics of populations,* Vol. 2., *The theory of gene frequencies.* Chicago: University of Chicago Press.

Young, A., Boyle, T., and Brown, T. (1996). The population genetic consequences of habitat fragmentation for plants. *Trends in Ecology and Evolution* 11, 413-418.

Young, A.G., Brown, A.H.D., Murray, B.G., Thrall, P.H., and Miller, C.H. (2000). Genetic, erosion, restricted mating and reduced viability in fragmented populations of the endangered grassland herb *Rutidosis leptorrhynchoides.* In *Genetics, demography and viability of fragmented populations* (pp. 335-360), Young, A.G. and Clarke G.M. (eds.). Cambridge: Cambridge University Press.

Young, A.G., Brown, A.H.D., and Zich, F.C. (1999). Genetic structure of fragmented populations of the endangered grassland daisy *Rutidosis leptorrhynchoides. Conservation Biology* 13, 256-265.

Chapter 9

Molecular Analysis
of Plant Genetic Resources

Glenn Bryan

RATIONALE FOR MOLECULAR ANALYSIS
OF PLANT GENETIC RESOURCES

The advent of plant molecular genetics in the late 1980s triggered a rapid increase in the number of studies in which plant germplasm was analyzed by using molecular genetic markers. These investigations supplement the conventional descriptions of gene bank and other plant material based on ecogeographic or "passport" information and on morphological characteristics. Molecular studies of crop plant germplasm have varied widely in the types of germplasm studied, the types of marker used, and the biological questions investigated.

There are essentially three types of investigations of plant genetic resources: the measurement and description of diversity, the development of effective methods of conservation, and the identification of material required by users, such as plant breeders. Molecular genetic markers can play a large role in all of these types of study. For example, they have been useful in gathering information on crop taxonomy and evolution, on the geographical and ecological aspects of genetic diversity, and on the processes that have given rise to observed patterns of variation. They have been used for gene bank management, for example, to identify duplicate accessions, to test accession "integrity," and to inform the identification of material for core collections. These activities are of critical importance owing to the high costs of maintaining and evaluating plant germplasm collections.

Molecular genetics studies have helped resolve issues of the phylogenetic or geographic origins of groups of plant species or of

doi:10.1300/5546_09

131

genomes present in polyploids, questions once thought to be of academic interest only but are now increasingly significant for understanding gene bank material. Germplasm collections are important reservoirs of as yet mostly untapped genes for the future production of plant cultivars. The availability of ever-changing molecular, biochemical, and metabolic assay technologies is generating a paradigm shift toward the assessment of biological or "functional" diversity as opposed to the use of "functionless" markers. Is this a valid approach? Usage of plant germplasm in crop improvement programs is heavily determined by particular phenotypic attributes, such as resistances to particular biotic or abiotic stresses. As advances in plant genomics provide more knowledge of the genes responsible for such traits, it seems wise to target investigations toward such genes. However, Tanskley and McCouch (1997) have suggested that the optimal rationale for finding novel alleles in exotic plant germplasm that significantly influence traits is to select accessions that show maximal dissimilarity to adapted cultivars of the crop itself, regardless of whether the diverged accessions appear to harbor useful variation for the targeted traits.

DNA fingerprinting techniques provide levels of germplasm identification and quantitative estimates of genetic diversity that are impossible to achieve with more plastic phenotypic descriptors. In general, molecular-based studies tend to support previous classifications based on other types of descriptors. There appears to be almost universal enthusiasm for their use, and it would be particularly attractive if molecular data could be combined with information from other sources. In this chapter I present my own views regarding the use of molecular markers to analyze plant germplasm, with reference to the recent literature, and try to provide insights as to where this very important area of research is likely to go next.

MOLECULAR MARKER TECHNIQUES AVAILABLE FOR THE ANALYSIS OF PLANT GENETIC RESOURCES

Marker technologies vary greatly in cost, reliability, ease of development, ease of deployment, multiplex ratios, efficiency, etc. The following discussion is a review of the various types of marker analysis used to date.

The first molecular technique was restriction fragment length polymorphism (RFLP). It is based on the variation between genotypes in DNA fragment lengths generated by restriction endonuclease digestion of their DNA (Lander and Botstein, 1986). RFLPs generally have a codominant mode of inheritance, allowing characterization of multiple alleles at a single locus. Most of the early plant genetic maps were based on use of RFLP, but, given the inherently low "multiplex ratio" (essentially one locus assayed per hybridization experiment) obtainable, this method is not ideally suited to analysis of larger sets of plant germplasm. However, RFLP analysis was used in some of the early studies of plant diversity (Brummer et al., 1991; Kochert et al., 1991).

Many RFLP probes have been sequenced and converted to polymerase chain reaction (PCR)-based "sequence-tagged sites" (STS). These sites are most commonly assayed for genetic polymorphism by restriction enzyme digestion of PCR products with frequently cutting restriction enzymes, generating cleaved amplified product (CAP) markers. These too have been used for measuring DNA polymorphism among sets of plant germplasm (Talbert et al., 1994).

Randomly amplified polymorphic DNA (RAPD) (Williams et al., 1990) became a prominent PCR-based marker system largely because it required no sequence information, making it very appealing to the majority of researchers working on plants with little or no sequence information. RAPD generates multilocus profiles comprising dominant markers. This technique is sensitive to changes in assay conditions, giving rise to concerns about their repeatability and transferability between laboratories. Some studies have addressed these problems, and radical improvements in reproducibility have been reported (Sobral and Honeycutt, 1993). Despite the criticisms, its technical simplicity and the small quantity of DNA required has made it useful for diversity and linkage analysis on a massive scale, particularly for "orphan" species for which few other techniques are available.

Simple sequence repeats (SSRs), also called *microsatellite* repeats, are stretches of short tandemly repeated motifs, generally between one and six nucleotides in length, and are dispersed throughout the genomes of most eukaryotes (Tautz and Renz, 1984). SSRs are characterized by extensive site-specific polymorphism between genotypes at a locus, generally due to differing numbers of repeat units.

The extensive length polymorphism observed at SSR loci has made them a very popular marker type for linkage mapping and genetic diversity studies. Length polymorphism at SSR loci is assayed by PCR, using primers designed for sequences flanking the repeat motif. SSR polymorphisms are generally codominant. The high level of allelic diversity at SSR loci in plants has been confirmed in many studies (Maroof et al., 1994), and large numbers of SSRs are now available for many plant species.

Several facile techniques for the enrichment of small insert genomic libraries for SSR motifs have been developed such that their isolation in large numbers is no longer problematic. Cardle et al. (2000) report that SSRs can be discovered at high frequency by partially sequencing large insert genomic clones. This approach lends itself to a targeted strategy for the development of SSR markers, impossible using a random cloning strategy, whereby SSRs can be developed from a bacterial artificial chromosome (BAC) arising from any chosen genetic location. It also allows the isolation of the most prevalent class of plant SSR (i.e., [AT]n), which are virtually impossible to identify by existing hybridization-based methods. More recently, sets of mononucleotide SSRs that target the chloroplast genome have been reported (Powell et al., 1995).

The amplified fragment length polymorphism (AFLP) technique is based on selective PCR amplification of restriction fragments from a complete digest of genomic DNA (Vos et al., 1995). Like RAPD, AFLP requires no prior sequence information. It also produces considerably more markers per assay and is inherently more robust. For these reasons, AFLP has generally replaced RAPD as a fingerprinting method.

The technique requires digestion of the DNA with two restriction endonucleases, with subsequent amplification of a subset of the restriction fragments using a modified PCR procedure. Products are visualized either by autoradiography (if radiolabeling one of the PCR primers is used) or by fluorescence detection on an automated sequencer (requiring that one primer be labeled with a fluorophore). Typically, 50 to 100 fragments (of 50 to 500 nucleotides) are detected, although this number is affected by factors including the size and complexity of the genome under examination, the number of selective nucleotides employed, and the endonucleases used. Polymorphism is generally detected as presence or absence of amplified fragments; thus, the majority of AFLPs are dominant in their mode of inheritance. The "two-state" nature of RAPD and AFLP markers somewhat limits their utility, notably in phylogenetic studies. The use of sequence motif information to develop AFLP-like systems targeting particular classes of sequences, such as retrotransposons, has been reported (Waugh et al., 1997), but space limitations prohibit further discussion here.

The knowledge that single-base substitutions represent the most frequent class of DNA sequence polymorphisms has been used to develop single nucleotide polymorphisms (SNPs) as a new marker type for linkage and diversity analysis in plants (Rafalski, 2002). SNPs are abundant in most plants and comprise biallelic markers that can be assayed using a wide array of detection platforms (see review by Gut, 2001), many of which require significant financial outlay at the outset. The cost of deploying SNPs at high throughput is a serious barrier to their widespread use, although the increasing availability of inexpensive detection methods (e.g., Ye et al., 2001) will eventually allow less sophisticated and well-financed laboratories to use SNPs. SNPs, SNP haplotypes, and other sequence-based approaches are likely to bring plant diversity studies closer to those of taxonomists, who have always advocated use of sequence data in favor of essentially "phenotypic" marker data. The possibility of a shift toward DNA sequence-based analytical methods for the measurement of plant genetic diversity is discussed in greater detail later in this chapter.

Diversity array technology employs a solid-state platform method to detect DNA polymorphism. It is low cost, has high throughput, and can be used for organisms with no available sequence information (Damian et al., 2001). Array technology is also available for assaying SNP data, but very few plants have sufficient SNPs to warrant the production of a microarray at present. Doubtless this will change, notably for the models and major crop plants, but for now we are limited to the analysis of DNA fragments or sequences for most plants.

TRENDS IN THE USE
OF MOLECULAR MARKERS

In recent years RFLP and RAPD have been gradually replaced by SSR and AFLP marker technologies for most genetic diversity studies. RAPD is losing popularity, partly due to problems of repro-ducibility and to a lack of information about the orthology of marker alleles across diverse material. It is also increasingly apparent that RAPD is not very efficient in the detection of polymorphisms despite its perceived low cost. It is remarkable that so many marker-based plant diversity studies, several hundred at least, were based on the use of RAPD, illustrating the very high uptake of this method particularly in smaller laboratories with limited access to advanced technology. SSRs can be difficult to deploy, especially across sets of very diverse plant material, and both AFLPs and SSRs are still predominantly used with radioisotopes and/or polyacrylamide gel electrophoresis. This will most likely change within the next few years, as automated sequencers become universally accessible and the cost of fluores-cently labeled primers decreases. Despite new enrichment proce-dures for the isolation of SSRs, their isolation in very large numbers remains problematic. There is generally a high attrition rate—only relatively few primer pairs developed generate reliable markers. SNPs are becoming increasingly adopted and will no doubt be used heavily in future crop species studies.

A major problem with almost all marker studies performed to date, and notably those employing dominant marker systems (e.g., AFLP and RAPD), is that they represent "closed" data sets that cannot be augmented in the future. There are very few marker-based studies of plant germplasm to which it has been possible to add additional markers or accessions. This is largely due to the vagaries of data scor-ing and annotation and to the use of different marker types and primer sets by different laboratories. There are very few sets of "standard" genotyping markers as yet adopted by the research community, al-though the barley SSR markers proposed by Macaulay et al. (2001) have potential.

LEVELS OF POLYMORPHISM AND CONGRUENCE OF DIFFERENT TYPES OF MOLECULAR DATA

The different PCR marker types described detect varying levels of genetic polymorphism. In short, it is clear that SSRs are the most polymorphic markers, followed by AFLP, which have the advantage of detecting large numbers of loci per gel track. RAPD is less poly-morphic than either and detects fewer loci than AFLP.

All molecular marker types have some sort of bias as to what fraction of the genome under study that they target. SSRs target particular repeat motifs, RAPDs target short inverted repeats, and AFLPs target particular combinations of restriction sites flanked by particular selective bases. Important studies have been performed to establish to what extent different marker types provide the same classification of plant germplasm. Notable are the comparative studies for soybean (Powell et al. 1996), potato (Milbourne et al., 1997), barley (Russell et al., 1997), rice (Virk et al., 2000), plantain (Ude et al., 2003), and garlic (Ipek et al., 2003). These studies have shown that data obtained using dominant marker systems such as RAPD and AFLP tend to concur much better than those obtained with SSR or isozymes. Tohme et al. (1996) assert that "AFLP analysis allows the production of a large amount of data in a short period, thus permitting greater insights into the genetic structure of wild beans than had been possible with other methods of analysis" (p. 1375).

Another question is whether it is preferable to use mapped markers for the measurement of genetic diversity. Virk et al. (2000) used rice as a model to examine this question and concluded that as long as the types of markers used are known to be widely distributed across the genome, the use of unmapped markers is not necessarily a disadvantage. Moreover, they postulate that if markers mapped on a cross between closely related parents were used, the markers may show a bias toward particular segments of the genome and thus give misleading information on patterns of diversity when applied to a wider collection of germplasm. Use of SNP markers whose locations are known a priori, based on their derivation from previously mapped RFLPs or genes, could help ensure that the markers are well distributed across the genome.

MOLECULAR MARKER ANALYSIS OF COLLECTIONS OF PLANT GENETIC RESOURCES

Major objectives of the numerous marker-based studies performed to date have been to identify duplicate accessions and to define "core collections." Both are somewhat controversial issues. In this section, I address these issues with some examples of markers used in for activities. Wherever possible, I mention studies that compare results obtained in molecular studies with those obtained by "conventional" approaches.

It is difficult to define exactly what represents a pair of duplicate accessions, particularly for outbreeding species, where accessions will undoubtedly contain significant heterozygosity. Potential duplicates have often been identified based on "passport" data, morphology, etc., as well as through the use of markers. However, for marker-based studies the situation is akin to that of use of markers for varietal identification or for stability following long-term tissue culture propagation. How many identical marker scores are needed to prove identity? How similar do two or more accessions have to be before we can say that they are the same? As always, it is easier to show that two entities are different than that they are identical. However, studies have shown that marker data can be very useful in the identification of duplicate or near-duplicate accessions of germplasm.

For example, Virk, Newbury, et al. (1995), in a well-designed study, used RAPD to examine accessions of *Oryza sativa* from IRRI that included known and suspected duplicates and very closely related germplasm. They described procedures for discriminating these categories of accessions that could also be used on a more general basis for identifying duplicates in gene banks. Baum et al. (1997) analyzed the genetic diversity in 88 genotypes from 20 populations of wild barley (*Hordeum spontaneum* C. Koch) from Israel, Turkey, and Iran using RAPD and found very few duplicate patterns. Perera et al. (1998) used AFLP profiling of Sri Lankan coconut palms to identify putative duplicate accessions for management of coconut germplasm and for optimizing choice of genetically divergent parents for crossing. Cao et al. (1998) used RAPD data to find several potential duplicates among accessions of spelta and macha wheats. Dansi et al. (2000) used RAPD data to discriminate among late maturing cultivars of Guinea yams (*Dioscorea cayenensis/D. rotundata* complex) that could not be separated using isozyme markers. Putative duplicates and cultivar misclassifications were identified, and morphologically distinct cultivars were observed as being very close using marker data, suggesting that morphological data should not be ignored in such studies.

Fregene et al. (2000) performed AFLP analysis of African cassava (*Manihot esculenta* Crantz) germplasm and found a considerable number of duplicates, suggesting significant redundancy. Hokanson et al. (2001) found that a set of SSRs were not useful in identifying genetic relationships among a diverse collection of *Malus* accessions, with the majority of accessions not clustering in ways concordant with taxonomic information and/or geographic origin. They also detected identical accessions in the collection that were previously considered to be unique. This study highlights a frequent criticism of SSR markers—that they are not reliable in studies that operate above the species level, although this is not a universal finding. Johnson et al. (2002) used 86 RAPD markers with 17 agronomic descriptors to analyze 228 accessions of *Poa pratensis* germplasm, failing to find any duplicates and concluding that unique genotypes were generally underrepresented in the collection. These authors observed a significant negative correlation between distance matrices based on RAPD and agronomic-based data and concluded that both molecular and agronomic characterizations were needed to assess overall diversity.

McGregor et al. (2002) examined a large collection of 314 wild potato accessions of the series Acaulia using two AFLP primer pairs to investigate the extent of redundancy and the distribution of diversity across the collection area. Misclassifications were identified, and 46 clones showed AFLP patterns identical to other accessions. Addition of data using a third primer pair reduced the number of suspected duplicates to fifteen. These authors concluded that AFLP analysis is an efficient method to verify taxonomic classification and to identify redundancies in wild germplasm of the series Acaulia.

Lund et al. (2003) performed an extremely detailed study using 35 SSR primer pairs covering the entire barley genome to find 36 potential duplicate groups among 174 "repatriated" accessions of barley. They suggest a "threshold average genetic distance value" for declaring two accessions to be different in a Nordic barley collection based on empirical analyses of distances within and between homogeneous groups of accessions.

Ipek et al. (2003) used AFLPs to examine 45 garlic (*Allium sativum* L.) clones and some near relatives, following their observation that isozymes and RAPD data generally agreed with morphological observations but sometimes failed to discriminate clones. Three AFLP primer combinations generated 183 polymorphic fragments, and identical AFLP profiles were observed, suggesting the presence of duplicates. A key problem with this type of study is that there is no a priori expectation of allele frequency in a sample (unlike linkage studies), so it is difficult to evaluate the effects of fragment scoring errors on the ability to detect duplicates in a sample.

The subject of core collections has been addressed by many studies. Ronning and Schnell (1994) suggested using allozyme and morphological data to establish a core collection of clones in order to maintain a cocoa germplasm collection. Lerceteau et al. (1997) evaluated the levels of genetic variability among cocoa accessions using RAPD and RFLP markers and suggested the use of their results for the selection of a cocoa tree core collection. Colombo et al. (1998) examined the genetic diversity of 31 Brazilian cassava clones, using RAPD and botanical descriptors. The two data sets revealed some identical relationships among the cultivars. However, the botanical descriptors were much less efficient at differentiating between pairs of cultivars than the molecular data, demonstrating the relative power of markers over botanical descriptors in studying genetic diversity, identifying duplicates, and validating or improving a core collection.

Ghislain et al. (1999) used RAPD markers, in concert with other data types, to select a core collection representing the biodiversity of Andean *Solanum phureja* potato germplasm. They also developed a protocol suitable for large germplasm collection genotyping, which was successful in identifying putative duplicates. Rebourg et al. (1999) examined RFLP diversity within and among 65 maize populations from various origins and concluded that classification of maize populations based on molecular markers should be helpful for genetic resources conservation and management through the definition of a representative subcollection (core collection). Chavarriaga-Aguirre et al. (1999) used a variety of marker types, including SSRs and AFLPs, to measure genetic diversity and redundancy in a large cassava core collection. A small number (1.34 percent) of potential duplicates were identified, and the authors suggest that traditional methods for selection of the core were highly successful in identifying unique genotypes and that future selections could employ DNA-based markers.

Dje et al. (2000) used microsatellite markers to quantify genetic diversity within and among sorghum accessions from the world germplasm collection. Their results were in good agreement with those obtained previously with allozyme markers, and the authors showed that microsatellite data are useful in identifying individual accessions with a high relative contribution to the overall allelic diversity of the collection. Grenier et al. (2000) performed a molecular marker-based assessment of genetic diversity in three core subsets selected from the

International Crops Research Institute for Semi-Arid Tropics sorghum collection using random and nonrandom sampling procedures. Polymorphism at 15 microsatellite loci was examined in each core, and the levels of allelic richness, average genetic diversity, and allelic frequencies were comparable for the three subsets. Moreover, a high percentage of rare alleles (71 to 76 percent of 206 total rare alleles) was maintained in the three subsets.

Carvalho and Schaal (2001) examined genetic diversity in a cassava germplasm collection in Brazil using SSR-based and RAPD markers. They confirm the close relationship of cassava to some other wild species. They also show that the relationships of accessions based on agronomic traits are not fully congruent with those revealed with RAPD markers and that the genetic diversity of the Brazilian cassava collection is not fully represented in Centro Internacional de Agricultura Tropical's World Core Collection. Dhanaraj et al. (2002) used RAPD to estimate the diversity among 90 cashew accessions from the National Cashew Gene Bank, identifying a core collection that represents the same diversity as the entire population. Dahlberg et al. (2002) compared levels of variation among 94 sorghum germplasm accessions using seed morphology and RAPD measurements. Their objective was to compare clusters developed from agronomic descriptors with groupings based on RAPD fingerprinting in an attempt to identify one approach using agronomic descriptors that would most closely approximate the groupings produced by RAPD markers. Put simply, their results suggested that none of four clustering methods used for agronomic descriptors closely approximated the groupings produced by RAPD markers. Ude et al. (2003) also stress the high relative value of AFLP markers, over RAPD, in the evaluation of a plantain core collection. The results of all of these studies underscore the need for further research into the techniques used to develop core collections.

TOWARD UNIVERSAL OR STANDARDIZED MARKER DATA SETS FOR PLANT GERMPLASM

The aforementioned studies show that markers can be more effective than previously used identifiers to quantify genetic relationships among germplasm. They are also helpful for the identification of potential duplicate accessions, selection of core collections, and prioritization of material for further use. This is not to say that conventional descriptors should be ignored; it is clear that they still have their uses. What is evident to this author is the lack of any major generality emerging from these studies and the absence of comparability between them. One of the few consistent observations is that SSR and AFLP markers are being increasingly used to evaluate gene bank material. The different data types and uses are summarized in Figure 9.1.

If fragment or allele sizes could be measured without error among different laboratories, then molecular marker data would be universally relevant to the research community for any particular plant by establishing marker data sets of general utility. For well-described single locus marker systems such as SSRs, this is largely true—alleles are sized to the basepair, so it is possible to establish, for example, a barley database containing data on 500 barley varieties at SSR loci (Macaulay et al., 2001). However, the establishment of the "Scottish Crop Research Institute (SCRI) barley SSR database" has not been

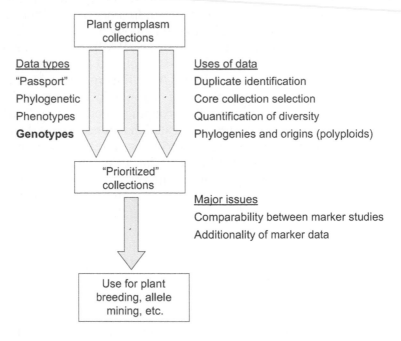

FIGURE 9.1. Summary of the Situation Regarding Usage of Molecular and Other Data for the Evaluation of Plant Germplasm Collections.

without technical problems, mostly caused by difficulties in scoring allele sizes unambiguously across different marker analysis platforms.

For dominant markers such as RAPD and AFLP, the situation is much more difficult. Markers are generally scored in such a way as to make them effectively closed to future augmentation. For AFLP in potato this perceived problem has been addressed by establishing an AFLP marker "catalogue" (Rouppe Van der Voort et al., 1997), with each individual locus being described by the AFLP primer sequences and a molecular weight. To date this approach has been used predominantly for the construction, by a consortium of laboratories, of a marker-dense linkage map in potato (e.g., Isidore et al., 2003) so that each partner can access the marker information from the whole map. If this type of approach could be extended to the analysis of plant genetic resources, it may be possible to develop marker sets for direct comparisons of different sets of plant germplasm or of the same collection over time to study genetic erosion and other phenomena.

A NEW PARADIGM:
A DNA SEQUENCE IS FOREVER!

There is a growing realization that marker fragment-based analyses of plant germplasm are not amenable to addition of data subsequent to the original analysis by either the same or different research groups. Aside from the problem of fragment labeling/nomenclature, there is the fact that marker data sets are, to some extent, defined by the subset of germplasm analyzed in any particular study and the type of method used to score data. The subsequent addition of more samples can alter the spectrum of loci observed in the overall study. This has significant implications for the storing and organization of data. For AFLPs, significant effort has been expended to develop sophisticated scoring software and a marker nomenclature system, which in principle allow the identification of any AFLP fragment by its primer extension and fragment size. This enables comparison of AFLP profiles between laboratories and the addition of data sets accrued at different times. Despite this advance, there are still significant problems with combining data sets that pose major obstacles to establishing marker databases that can be added to in the future. The situation is complicated even further by the ever-increasing array of technology platforms available for marker-based analyses of plant germplasm. Automated systems (e.g., ABI, LICOR, Beckman, etc.) introduce yet another level of variability.

A possible solution is to perform all future analyses of plant germplasm using a DNA sequence-based approach. Sequence data are "absolute" and do not depend on any particular technology platform. They can be supplemented or verified at any future time by the same or other investigators. They are also inherently phylogenetically more informative than marker data. However, the use of sequence data has its own set of challenges. It remains relatively expensive, and there are questions about which gene or genes to sequence (see Schaal et al., 2003): Which genes have enough sequence diversity for within-species studies? How do we deal with outbreeders and polyploids? How do we know that genes are single copy and that we are not sequencing paralogous genes? How many genes do we need to sequence to avoid simply producing "gene trees"? The replacement of molecular markers with sequence-based approaches is fraught with serious challenges, both technical and conceptual.

Recent studies performed using the sequences of chloroplast and single-copy nuclear genes to study the evolution of crop plant taxa (e.g., Cronn et al., 2002). are encouraging and require further consideration and discussion. The establishment of gene sequence databases for related groups of plants, both within and between species, could revolutionize the analysis of plant genetic resources. The identification of sets (15 to 25) of single-copy PCR amplicons would be a significant first step in this process. Alternatively, a move toward the use of SNP markers or haplotypes would satisfy the need for a sequence-based approach to the analysis of plant germplasm, and this area is rapidly growing.

CONCLUSION

Molecular markers have had a profound effect on our ability to catalogue and describe plant germplasm and to prioritize its use in plant breeding programs. Morever, they represent an extremely cost-efficient method for generating such information, especially compared with the massive costs of maintaining large plant germplasm collections. However, certain changes in approach need to be made if molecular data sets of individual plant collections are to become generally applicable, comparable, or extendable, and these should include a move away from fragment-based marker systems toward the use of DNA sequence information.

REFERENCES

Baum, B.R., Nevo, E., Johnson, D.A., and Beiles, A. (1997). Genetic diversity in wild barley (*Hordeum spontaneum* C Koch) in the Near East: A molecular analysis using random amplified polymorphic DNA (RAPD) markers. *Genetic Resources and Crop Evolution* 44, 147-157.

Brummer, E.C., Kochert, G., and Bouton, J.H. (1991). RFLP variation in diploid and tetraploid alfalfa. *Theoretical and Applied Genetics* 83, 89-96.

Cao, W.G., Hucl, P., Scoles, G., and Chibbar, R.N. (1998). Genetic diversity within spelta and macha wheats based on RAPD analysis. *Euphytica* 104, 181-189.

Cardle, L., Ramsay, L., Milbourne, D., Macaulay, M., Marshall, D., and Waugh, R. (2000). Computational and experimental characterization of physically clustered simple sequence repeats in plants. *Genetics* 156, 847-854.

Carvalho, L. and Schaal, B.A. (2001). Assessing genetic diversity in the cassava (*Manihot esculenta* Crantz) germplasm collection in Brazil using PCR-based markers. *Euphytica* 120, 133-142.

Chavarriaga-Aguirre, P., Maya, M.M., Tohme, J., Duque, M.C., Iglesias, C., Bonierbale, M.W., Kresovich, S., and Kochert, G. (1999). Using microsatellites, isozymes and AFLPs to evaluate genetic diversity and redundancy in the cassava core collection and to assess the usefulness of DNA-based markers to maintain germplasm collections. *Molecular Breeding* 5, 263-273.

Colombo, C., Second, G., Valle, T.L., and Charrier, A. (1998). Genetic diversity characterization of cassava cultivars (*Manihot esculenta* Crantz): I. RAPD markers. *Genetics and Molecular Biology* 21, 105-113.

Cronn, R.C., Small, R.L., Haselkorn, T., and Wendel, J.F. (2002). Rapid diversification of the cotton genus (*Gossypium*: Malvaceae) revealed by analysis of sixteen nuclear and chloroplast genes. *American Journal of Botany* 89, 707-725.

Dahlberg, J.A., Zhang, X., Hart, G.E., and Mullet, J.E. (2002). Comparative assessment of variation among sorghum germplasm accessions using seed morphology and RAPD measurements. *Crop Science* 42, 291-296.

Damian, J., Peng, K., Feinstein, D., and Kilian, A. (2001). Diversity arrays: A solid state technology for sequence information independent genotyping. *Nucleic Acids Research* 29(4), 25.

Dansi, A., Mignouna, H.D., Zoundjihekpon, J., Sangare, A., Ahoussou, N., and Asiedu, R. (2000). Identification of some Benin Republic's guinea yam (*Dioscorea cayenensis Dioscorea rotundata* complex) cultivars using randomly amplified polymorphic DNA. *Genetic Resources and Crop Evolution* 47, 619-625.

Dhanaraj, A.L., Rao, E.V.V., Swamy, K.R.M., Bhat, M.G., Prasad, D.T., and
    Sondur, S.N., (2002). Using RAPDs to assess the diversity in Indian cashew
    (*Anacardium occidentale* L.) germplasm. *Journal of Horticultural Science &
    Biotechnology* 77, 41-47.
Dje, Y., Heuertz, M., Lefebvre, C., and Vekemans, X. (2000). Assessment of ge-
    netic diversity within and among germplasm accessions in cultivated sorghum
    using microsatellite markers. *Theoretical and Applied Genetics* 100, 918-925.
Fregene, M., Bernal, A., Duque, M., Dixon, A., and Tohme, J. (2000). AFLP analy-
    sis of African cassava (*Manihot esculenta* Crantz) germplasm resistant to the
    cassava mosaic disease (CMD). *Theoretical and Applied Genetics* 100, 678-685.
Ghislain, M., Zhang, E., Fajardo, D., Huaman, Z., and Hijmans, R.J. (1999).
    Marker-assisted sampling of the cultivated Andean potato *Solanum phureja* col-
    lection using RAPD markers. *Genetic Resources and Crop Evolution* 46, 547-
    555.
Grenier, C., Deu, M., Kresovich, S., Bramel-Cox, P.J., and Hamon, P. (2000). As-
    sessment of genetic diversity in three subsets constituted from the ICRISAT sor-
    ghum collection using random vs. non-random sampling procedures: B. Using
    molecular markers. *Theoretical and Applied Genetics* 101, 197-202.
Gut, I.G. (2001). Automation in genotyping of single nucleotide polymorphisms.
    Human Mutation 17, 475-492.
Hokanson, S.C., Lamboy, W.F., Szewc-Mcfadden, A.K., and Mcferson, J.R. (2001).
    Microsatellite (SSR) variation in a collection of *Malus* (apple) species and hy-
    brids. *Euphytica* 118, 281-294.
Ipek, M., Ipek, A., and Simon, W.P. (2003). Comparison of AFLPs, RAPD markers,
    and isozymes for diversity assessment of garlic and detection of putative dupli-
    cates in germplasm collections. *Journal of the American Society for Horticul-
    tural Science* 128, 246-252.
Isidore, I., Van Os, H., Andrzejewski, S., Bakker, J., Barrena, I., Bryan, G.J.,
    Caromel, B., van Eck, H., Ghareeb, B., de Jong, W., et al. (2003). Calculation of
    ultra-high density genetic linkage maps of heterozygous outbreeders: A model
    study of chromosome one of potato. *Genetics* 165, 2107-2116.
Johnson, R.C., Johnston, W.J., Golob, C.T., Nelson, M.C., and Soreng, R.J. (2002).
    Characterization of the USDA *Poa pratensis* collection using RAPD markers
    and agronomic descriptors. *Genetic Resources and Crop Evolution* 49, 349-361.
Kochert, G., Halward, T., Branch, W.D., and Simpson, C.E. (1991). RFLP variabil-
    ity in peanut (*Arachis-Hypogaea* L) cultivars and wild-species. *Theoretical and
    Applied Genetics* 81, 565-570.
Lander, E.S. and Botstein, D. (1986). Mapping complex genetic-traits in humans:
    New methods using a complete RFLP linkage map. *Cold Spring Harbor Sympo-
    sia on Quantitative Biology* 51, 49-62.
Lerceteau, E., Robert,T., Petiard, V., and Crouzillat, D. (1997). Evaluation of the
    extent of genetic variability among *Theobroma cacao* accessions using RAPD
    and RFLP markers. *Theoretical and Applied Genetics* 95, 10-19.

Lund, B., Ortiz, R., Skovgaard, I.M., Waugh, R., and Andersen, S.B. (2003). Analysis of potential duplicates in barley gene bank collections using re-sampling of microsatellite data. *Theoretical and Applied Genetics* 106, 1129-1138.

Macaulay, M., Ramsay, L., Powell, W., and Waugh, R. (2001). A representative, highly informative "genotyping set" of barley SSRs. *Theoretical and Applied Genetics* 102, 801-809.

Maroof, M.A.S., Biyashev, R.M., Yang, G.P., Zhang, Q., and Allard, R.W. (1994). Extraordinarily polymorphic microsatellite DNA in barley—Species-diversity, chromosomal locations, and population-dynamics. *Proceedings of the National Academy of Sciences of the United States of America* 91, 5466-5470.

McGregor, C.E., Van Treuren, R., Hoekstra, R., and Van Hintum, T.J.L. (2002). Analysis of the wild potato germplasm of the series Acaulia with AFLPs: Implications for ex situ conservation. *Theoretical and Applied Genetics* 104, 146-156.

Milbourne, D., Meyer, R., Bradshaw, J.E., Baird, E., Bonar, N., Provan, J., Powell, W., and Waugh, R. (1997). Comparison of PCR-based marker systems for the analysis of genetic relationships in cultivated potato. *Molecular Breeding* 3, 127-136.

Perera, L., Russell, J.R., Provan, J., Mcnicol, J.W., and Powell, W. (1998). Evaluating genetic relationships between indigenous coconut (*Cocos nucifera* L.) accessions from Sri Lanka by means of AFLP profiling. *Theoretical and Applied Genetics* 96, 545-550.

Powell, W., Morgante, M., Andre, C., Hanafey, M., Vogel, J., Tingey, S., and Rafalski, A. (1996). The comparison of RFLP, RAPD, AFLP and SSR (microsatellite) markers for germplasm analysis. *Molecular Breeding* 2, 225-238.

Powell, W., Morgante, M., Andre, C., McNicol, J.W., Machray, G.C., Doyle, J.J., Tingey, S.V., and Rafalski, J.A. (1995). Hypervariable microsatellites provide a general source of polymorphic DNA markers for the chloroplast genome. *Current Biology* 5, 1023-1029.

Rafalski, J.A. (2002). Novel genetic mapping tools in plants: SNPs and LD-based approaches. *Plant Science* 162, 329-333.

Rebourg, C., Dubreuil, P., and Charcosset, A. (1999). Genetic diversity among maize populations: Bulk RFLP analysis of 65 accessions. *Maydica* 44, 237-249.

Ronning, C.M. and Schnell, R.J. (1994). Allozyme diversity in a germplasm collection of *Theobroma-cacao* L. *Journal of Heredity* 85, 291-295.

Rouppe van der Voort, J.N.A.M., Van Eck, H.J., Draaistra, J., Van Zandvoort, P.M., Jacobsen, E., and Bakker, J. (1998). An online catalogue of AFLP markers covering the potato genome. *Molecular Breeding* 4, 73-77.

Russell, J.R., Fuller, J.D., Macaulay, M., Hatz, B.G., Jahoor, A., Powell, W., and Waugh, R. (1997). Direct comparison of levels of genetic variation among barley accessions detected by RFLPs, AFLPs, SSRs and RAPDs. *Theoretical and Applied Genetics* 95, 714-722.

Schaal, B.A., Gaskin, J.F., and Caicedo, A.L. (2003). Phylogeography, haplotype trees, and invasive plant species. *Journal of Heredity* 94, 197-204.

Sobral, B.W.S. and Honeycutt, R.J. (1993). High output genetic-mapping of poly-ploids using Pcr-generated markers. *Theoretical and Applied Genetics* 86, 105-112.

Talbert, L.E., Blake, N.K., Chee, P.W., Blake, T.K., and Magyar, G.M. (1994). Evaluation of sequence-tagged-site Pcr products as molecular markers in wheat. *Theoretical and Applied Genetics* 87, 789-794.

Tanksley, S.D. and McCouch, S.R. (1997). Seed banks and molecular maps: Unlocking genetic potential from the wild. *Science* 277, 1063-1066.

Tautz, D. and Renz, M. (1984). Simple sequences are ubiquitous repetitive components of eukaryotic genomes. *Nucleic Acids Research* 12, 4127-4138.

Tohme, J.D., Gonzalez, O., Beebe, S., and Duque, M.C. (1996). AFLP analysis of gene pools of a wild bean core collection. *Crop Science* 36, 1375-1384.

Ude, G., Pillay, M., Ogundiwin, E., and Tenkouano, A. (2003). Genetic diversity in an African plantain core collection using AFLP and RAPD markers. *Theoretical and Applied Genetics* 107, 248-255.

Virk, P.S., Ford-Lloyd, B.V., Jackson, M.T., and Newbury, H.J. (1995). Use of RAPD for the study of diversity within plant germplasm collections. *Heredity* 74, 170-179.

Virk, P.S., Newbury, H.J., Jackson, M.T., and Ford-Lloyd, B.V. (1995). The identification of duplicate accessions within a rice germplasm collection using RAPD analysis. *Theoretical and Applied Genetics* 90, 1049-1055.

Virk, P.S., Newbury, H.J., Jackson, M.T., and Ford-Lloyd, B.V. (2000). Are mapped markers more useful for assessing genetic diversity? *Theoretical and Applied Genetics* 100, 607-613.

Vos, P., Hogers, R., Bleeker, M., Reijans, M., Van de lee, T., Hornes, M., Frijters, A., Pot, J., Peleman, M., Kuiper, M., and Zabeau, M. (1995). AFLP—a new technique for DNA-fingerprinting. *Nucleic Acids Research* 23, 4407-4414.

Waugh, R., McLean, K., Flavell, A.J., Pearce, S.R., Kumar, A., Thomas, B.B.T., and Powell, W. (1997). Genetic distribution of Bare-1-like retrotransposable elements in the barley genome revealed by sequence-specific amplification polymorphisms (S-SAP). *Molecular & General Genetics* 253, 687-694.

Williams, J.G.K., Kubelik, A.R., Livak, K.J., Rafalski, J.A., and Tingey, S.V. (1990). DNA polymorphisms amplified by arbitrary primers are useful as genetic-markers. *Nucleic Acids Research* 18, 6531-6535.

Ye, S., Dhillon, S., Ke, X., Collins, A.R., and Day, I.N.M. (2001). An efficient procedure for genotyping single nucleotide polymorphisms. Nucleic Acids Research 29(17), 88.

Chapter 10

Analysis of Nuclear, Mitochondrial, and Chloroplast Genomes in Plant Conservation

C. Q. Sun

INTRODUCTION

The nucleus, mitochondria, and chloroplast are the sites of genetic information in plant cells. They contain their own genomes and can be autonomously replicated or independent in inheritance. Nuclear genetic information is inherited from both parents. Mitochondrial and chloroplast information is usually inherited maternally; some exceptions are *Oenothera,* alfalfa, and conifers (Neale and Oerofe, 1989). Molecular markers are powerful tools for assessing genetic differentiation of the three plant genomes. Knowledge about the genetic differentiation of nuclear, mitochondrial, and chloroplast genomes enables us to clarify the biosystematic relationship among different species.

ANALYSIS OF NUCLEAR GENOMES

The nuclear genomes of plants show tremendous variability in size. The flowering plants analyzed to date have genome sizes ranging from 0.1 pg to over 125 pg (Leitch et al., 1998). *Arabidopsis* has one of the smallest genomes in plants, 125Mbp (Arabidopsis Genome Initiative, 2000). Differences in genome size can mainly be attributed to varying proportions of repeated DNA sequences (Flavell,

doi:10.1300/5546_10

1980), although ploidy level is another factor that affects the size of genomes.

Abundant DNA markers from nuclear genomes, such as RFLP (restriction fragment length polymorphism), RAPD (random amplified polymorphic DNA), AFLP (amplified fragment length polymorphism), and SSR (simple sequence repeat), are powerful tools with which to study variations of plant nuclear genomes. Molecular data are used to assess the evolutionary relatedness of plant species. The similarity of each pair of samples can be measured by calculating the number of common DNA markers (Nei and Li, 1979; Nei, 1987). The similarity data can then be used in a cluster analysis to develop dendrograms, which show the molecular relatedness of the species.

There are many reports of analysis of DNA polymorphism and phylogenetic relationships among plant species using molecular markers. For rice, Wang et al. (1992) examined the RFLP of 93 accessions representing 21 species from the genus *Oryza* and concluded that the classification of *Oryza* species based on RFLP confirm previous classifications based on morphology, hybridization, and isozymes. Sun et al. (2002) analyzed the genetic diversity of the nuclear genomes of common wild rice (*Oryza rufipogon* Griff.) and cultivated rice (*O. sativa* L.). It was demonstrated that the Chinese common wild rice could be classified into three types, *indica*-like, *japonica*-like, and Chinese wild-specific types, while the South Asian common wild rice could be clustered into only wild-specific and *indica*-like types. There were distinguishing differences among the Chinese common wild rice and South and Southeast Asian common wild rice. The differences between *indica* and *japonica* subspecies of rice were also investigated using RFLP, RAPD, AFLP, and SSR markers (Wang and Tanksley, 1989; Zhang et al., 1992; Doi et al., 1995; Virk et al., 2000; Parsons et al., 1997; Zhu et al., 1998; Zhu et al., 2001; Chen et al., 2001). These analyses are very helpful in determining the origins and domestication of Asian cultivated rice. Similar analyses have been applied to other plant groups.

For example, variations in nuclear DNA helped define the phylogenetic relationships in *Brassica* and its related genera (Song et al., 1988, 1990). *Brassica* species can be divided into two evolutionary pathways: the *nigra* lineage and the *rapa/oleracea* lineage based on this evidence.

The 5S rRNA genes occur in tandom arrays in the nuclear genome with highly conserved regions and are separated by the nontranscribed intergenic spacer, which can vary in length or sequence between and even within species. Because the intergenic spacer evolves rapidly and is informative at the level of genus and species, it has often been chosen to study the phylogenetic relationships between and within genera (Sano and Sano, 1990; Baum and Appels, 1992; McIntyre et al., 1992; Moran et al., 1992; Udovicic et al., 1995).

The genetic diversity of germplasm of cultivated plants has also been studied frequently by nuclear genomic analysis. Sun et al. (2001) demonstrated that the genetic diversity of cultivated rice is lower than that of common wild rice. The number of polymorphic loci of cultivated rice is only three-quarters of that of common wild rice. The number of alleles is 60 percent and the number of genotypes is about 50 percent of that of common wild rice.

COMPARATIVE ANALYSIS OF THE GENOMES OF PLANTS

Comparative genomics is the study of the similarities and differences in the structure and function of hereditary information across taxa using molecular tools (Paterson et al., 2000). In the past, comparative mapping was widely used to study the genetic variations and phylogenetic relationships between species and subspecies. The similarities of the nuclear genomes between species can be evaluated by comparative mapping and sequences. Often the clones or sequence information used to identify markers in one species can be used to develop a genetic map for another related species. Once the two species have been mapped, the relative evolution of their genomes can be compared. Comparative mapping can identify chromosome inversions, translocations, and duplications that have occurred. Genetic factors can also be assessed by comparing map distances of genes and conserved gene order.

A study of the related grass species sorghum and maize revealed that many of the sorghum chromosomes contained regions from two of the maize chromosomes (Whitkus et al., 1992). This may be a result of the ancestral duplication of chromosomal material. Furthermore, duplicate genes have occurred to a greater extent in maize than in sorghum evolution. Detailed comparisons of these species at the megabase and sequence level are described by Bennetzen (2000).

The rice genome, with only 400 million DNA base pairs (Mb), is the smallest among the major cereal crops. Dense genetic maps have been constructed, and physical maps cover most of the genome. In addition, entire genomic sequences and expressed sequence tags (ESTs) are available. Therefore, the rice plant has been promoted as a model for cereal crops in comparative genomics. Based on the concept of conserved linkage segments, multiple alignments of chromosome maps are possible, and a comparative map including the genomes of foxtail millet, oats, pearl millet, maize, rice, sugarcane, sorghum, and Triticeae was developed (Gale and Devos, 1998a,b; Dunford et al., 1995). Fine-scale collinearity is maintained between rice and the Triticeae (Dunford et al., 1995). Moreover, the highly similar arrangement of linkage blocks in foxtail millet and rice demonstrate that these genomes have undergone few rearrangements since their divergence. The pearl millet genome, on the other hand, is highly rearranged relative to rice (Devos and Gale, 2000).

Comparative mapping has generally revealed collinear chromosomal segments in closely related plants, albeit of varying sizes. In some cases entire chromosomes show collinearity. For example, rice chromosome 9 is collinear with the consensus map for group 5 chromosomes of wheat, but a nonsyntenic region has been pinpointed (Foote et al., 1997). Collinearity was apparent between rice chromosome 10 and sorghum and maize based on the sequence of chromosome 10 (Rice Chromosome 10 Sequencing Consortium, 2003). Likewise, a detailed comparative study of the *Rpg1* genomic regions in rice provided evidence for a translocation that disrupts collinearity (Kilian et al., 1997).

It is important to investigate collinearity of genomes at the sequence level because it is difficult to assess to what extent genomes are conserved in two species based solely on the results of comparative genetic and physical mapping experiments. The entire genome of some model plants (rice, *Arabidopsis,* etc.) has been sequenced (Goff et al., 2002; Yu et al., 2002; Sasaki et al., 2002; Feng et al., 2002; Arabidopsis Genome Initiative, 2000). Large numbers of sequences of full-length complementary DNA clones of rice have been published (Kikuchi et al., 2003). Abundant databases of DNA sequences from different cultivars, subspecies, and species make it possible to compare the variances between species of plant at the single nucleotide polymorphism (SNP) level (Schmidt, 2002; Han and Xue, 2003). Bennetzen (2000) demonstrated that microstructure might not be as conserved as the gross chromosomal organization by comparative sequence analyses. Comparison of microcollinearity in the *sh2*-homologous locus of the maize, rice, and sorghum genomes indicated that in both rice and sorghum the *sh2* and *a1* genes are separated by ca.19 kb, whereas in maize the two genes are 140 kb apart (Chen et al., 1997). Sixty-four percent of cDNAs from *japonica* rice were homologous to *Arabidopsis* proteins (Kikuchi et al., 2003), and about 49.4 percent of predicted rice genes had a homolog in the *A. thaliana* genome (Yu et al., 2002)

ANALYSIS OF THE MITOCHONDRIAL GENOME

Plant mitochondrial genomes are the largest characterized mitochondrial genomes and the most variable in size. Sizes range from 200 kb in *Solanum tuberosum* to 2,000 kb in *Cucumis melo* (Eckenrode and Levings, 1986; Newton, 1988). Much of the genome size is due to the presence of noncoding sequence and of coding and noncoding segments of DNA that are present in multiple copies in the genome. Within the past ten years, many complete mtDNA sequences have been determined. The entire mitochondrial genome of rice (*Oryza sativa* L.) has been sequenced, and it was found to comprise 490,520 bp, with an average G+C content of 43.8 percent (Notsu et al., 2002). Turmel et al. (2002) determined the complete cpDNA (131,138bp) and mtDNA (56,574bp) sequences of the charophyte *Chaetosphaeridium globosum* (Colechaetales).

There have been many analysis of mitochondrial DNA variation within and among species in past two decades (Breiman, 1987; Ishii et al., 1993; Luo et al., 1995; Gray et al., 1999; Mackenzie and McIntosh, 1999). Ishii et al. (1993) reported that mitochondrial genomes of eight varieties of *O. sativa* and two varieties of *O. glaberrima* could be classified into five types by RFLP analysis. Sun et al. (2002) distinguished the genetic differentiation of mitochondrial genomes of *O. rufipogon* and *O. sativa* with 193 accessions. All these materials were clustered into five groups based on the RFLP data of mtDNA. Mitochondrial genomes of 75 cultivars of rice could be clustered into *indica* and *japonica* types, corresponding to *indica* and *japonica* sub-species, except for 'Bonsaj', a cultivar from Bangladesh. The ratio of *indica*- to *japonica*-type mitochondrial genomes is near 1:1, suggesting that the *indica-japonica* differentiation is also a major distinction in mitochondrial genomes of cultivated rice. In *O. rufipogon,* two other kinds of wild-type mitochondrial genome specific to *O. rufipogon* were detected besides the *indica*-like and the *japonica*-like types that are shared with cultivars.

Mitochondrial gene markers have also been a potentially powerful tool for the analysis of population differentiation. Wu et al. (1998) examined mtDNA polymorphism in pines derived from different populations with RFLP and concluded that genetic diversity within populations is much less than that among populations.

The mutation rate of the mitochondrial genome has been well investigated. Wolf et al. (1987) reported that the mutation rate of the mitochondrial DNA is lower than that of nuclear and chloroplast DNA. Sun et al. (2002) also revealed that the evolutionary variability of the nuclear genomes is higher than that of the mitochondrial genomes by comparison of the genetic distances of nuclear and mitochondrial genomes of common wild rice and cultivated rice. Palmer and Herbon (1988) also indicated that plant mitochondrial DNA evolves slowly in sequence although rapidly in structure. However, Ishii et al. (1993) concluded that the evolutionary variability of the mitochondrial and nuclear genomes were almost the same based on the genetic distances among the ten cultivars of *O. sativa* and *O. glaberrima.*

Analysis of the intron regions of mitochondrial genes is also useful in revealing relationships between species. Sanjur et al. (2002) investigated the phylogenetic relationships among six wild and six domesticated taxa of *Cucurbita* using an intron region from the mitochondrial *nad1* gene as a marker. Their study represents one of the first successful uses of a mtDNA gene in resolving inter- and intraspecific taxonomic relationships in angiosperms and yields several important insights into the origins of domesticated *Cucurbita.*

ANALYSIS OF THE CHLOROPLAST GENOME

The chloroplast genomes of plants are also circular molecules of double-stranded DNA. The complete sequences of the genomes of the chloroplasts of some species have been determined in the past decade. The chloroplast genome of rice is 134,525 bp in size (Hiratsuka et al., 1989). The genome of the chloroplasts of *Marchantia polymorpha* (a liverwort) contains 121,024 bp in a closed circle. The nucleotide sequence of the chloroplast (cp) DNA from maize *(Zea mays)* consists of 140,387 bp. The gene content and the relative positions of a total of 104 genes in the chloroplast genome of maize are identical with the cpDNA of the closely related species rice (Maier et al., 1995).

Analysis of cpDNA diversity has been used to determine taxonomy and to retrace phylogenetic relationships in higher plants (Timothy et al., 1979; Kemble and Shepard, 1984; Ichikawa et al., 1986; Breiman, 1987; Neale and Oerofe, 1989; Dally and Second, 1990; Laurent et al., 1993; Ishii et al., 1986, 1988, 1993; Luo et al., 1995; Demesure and Petit, 1995; Demesure et al., 1996; El Mousadik and Petit, 1996; Cipriani et al., 2003; Hadis et al., 2003). For rice, Ishii et al. (1986) and Dally and Second (1990) showed that there were two major types of cpDNA in Asian cultivated rice. A deletion of 69 bp sequence in the ORF100 (open reading frame 100) region was identified in cpDNA of *indica* varieties compared with *japonica* varieties (Ishii et al., 1986, 1988; Kanno et al., 1993). The deletion can be detected by polymerase chain reaction amplification (Kanno et al., 1993; Chen et al., 1993, 1994; Sun et al., 2002). It was demonstrated that most of the *japonica* varieties of rice showed *japonica*-type cpDNA, and most of the *indica* varieties showed *indica*-type cpDNA. However, a few *indica* cultivars showed *japonica*-type cpDNA, and some *japonica* cultivars are of *indica*-type cpDNA (Dally and Second, 1990; Sun et al., 2002).

The *indica-japonica* differentiation of chloroplast genome was also detected in common wild rice (*O. rufipogon* Griff.). Based on the *indica-japonica* differentiation of common wild rice, a "diphyletic hypothesis" for origin of cultivated rice was proposed (Ishii et al., 1988; Dally and Second, 1990; Sun et al., 2002).

Highly polymorphic SSRs were also discovered in the chloroplast genomes of plants, providing new strategies for high-resolution cytoplasmic population and phylogenetic analyses (Powell et al., 1995). Chloroplast genome SSR (cpSSR) studies reveal higher polymorphism than those using RFLP (Powell et al., 1995; Provan et al., 1998), thus allowing analysis of intraspecific variation in the chloroplast genomes of plants (Powell et al., 1995). Provan et al. (1999) analyzed the cpSSR in the genus *Hordeum*, which includes cultivated barley and its wild progenitor. A decrease in cytoplasmic diversity was observed between the wild progenitor and cultivated barley.

The chloroplast genome is conservative during the course of evolution. However, differentiation of chloroplast genomes within natural populations of common wild rice were detected (Huang et al., 1996). Two kinds of cpDNA, the *indica* and *japonica* types, were detected in different individuals within a natural population of common wild rice.

Transfer of Mitochondrial and Chloroplast Genes to Nuclear Genomes

Mitochondrial genomes are derived from the genome of a bacterial endosymbiont, with many genes having been lost or transferred to the nucleus early in mitochondrial evolution (reviewed by Gray, 1992; Gray et al., 1999). Berg and Kurland (2000) found that, under realistic conditions, the transfer from mitochondria to nuclei is inevitable for genes that can function equally well in the mitochondria and in the nucleus. Adams et al. (2002) demonstrated high and variable rates of mitochondrial gene loss and transfer to the nucleus during angiosperm evolution through a survey of the presence or absence of 40 mitochondrial protein genes in 280 genera of flowering plants.

Widespread horizontal transfer of mitochondrial genes in flowering plants was clarified by Bergthorsson et al. (2003). These results imply the existence of mechanisms for the delivery of DNA between unrelated plants and indicate that horizontal transfer is also a force in plant nuclear genomes.

Many genes have been lost from the chloroplast genome during plant and algal evolution. Most of these losses occurred in the murky interval between the original endosymbiosis of a *Cyanobacterium* (with perhaps 2,000 protein-coding genes) and the last common ancestor of all existing chloroplast genomes (Martin et al., 1998). Millen et al. (2001) analyzed the phylogenetic relationship of *infA*, which codes for translation initiation factor 1. Sequences and assessment of transit peptide homology indicate that the four nuclear *infA* genes are probably derived from four independent gene transfers from chloroplast to nuclear DNA during angiosperm evolution. It is also proposed that the gene has probably been transferred many more times, considering the many separate losses of *infA* from chloroplast DNA.

Comparison of the Differentiation Among Nuclear, Mitochondrial, and Chloroplast Genomes

The nuclear genome containing most genes presumably plays some role in regulating the relative abundance of various mtDNA and cpDNA configurations. However, the types of nuclear, mitochondrial, and chloroplast genomes were not consistent in rice accessions. Sun et al. (2002) showed that about one-quarter of all cultivars of rice were heterotypes. Almost all heterotype strains are landrace varieties. The proportion of homeotypes between nuclear and mitochondria, nuclear and chloroplast, and mitochondria and chloroplast in common wild rice was much lower than that in cultivated rice (Sun et al., 2002). This suggests that the coincidence of genetic differentiation among the three different classes of genome has increased with domestication. It also could be concluded that the differentiation of the three different classes of genome in wild species is more complicated than that in cultivated species.

Clustering Based on the Genetic Differentiation of the Three Different Classes of Genome

Analysis of genetic diversity and relationships between accessions is essential to evaluate and utilize germplasm. Classification of relationships based on molecular data should be more accurate than those developed using morphological traits and isozyme markers. However, classifications considering the genetic differentiation of the three classes of genomes may be more comprehensive in revealing the variations between species or types. Of the three genomes, the nuclear genome plays a principal role in the growth and development of plants. Thus it is necessary to classify the accessions into main groups based on the genetic differentiation of the nuclear genome and then subdivided based on cytotypes.

According to the genetic differentiation of nuclear genomes and considering the cytoplasmic type, cultivated rices were classified into two major groups and seven subgroups (Figure 10.1). However,

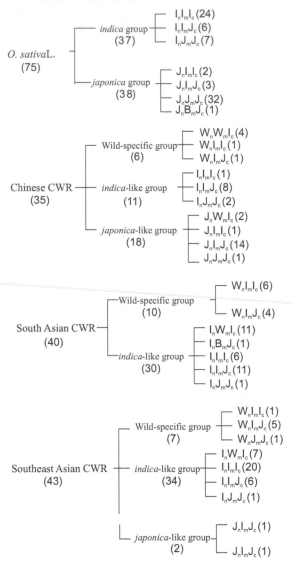

FIGURE 10.1. Classification of the Common Wild Rice (CWR) and Cultivated Rice Based on the *indica-japonica* Differentiation of the Three Genomes (*Source:* Sun et al., 2002, reprinted with kind permission of Springer Science and Business Media) *Note:* The numbers in parentheses are numbers of strains; W, I, J, and B stand for wild-specific type, *indica* or *indica*-like type, *japonica* or *japonica*-like type, and specific type for Bangladesh strains, respectively; n, m, and c stand for nuclear, mitochondrial, and chloroplast genomes, respectively.

common wild rices from China, South Asia, and Southeast Asia were classified into three groups and ten subgroups, two groups and seven subgroups, and three groups and nine subgroups, respectively (Figure 10.1). All the types observed in cultivated rice could be detected in common wild rice, but many types detected in the common wild rice could not be found in the cultivated rice (Sun et al., 2002). This demonstrates that the genomic type in wild species is much more diversified than that in cultivated species, and the genetic diversity has decreased in nuclear, mitochondrial, and chloroplast genomes during the course of evolution from wild to cultivated species.

Inference on the Origin and Domestication of Cultivated Plants Based on the Differentiation of the Three Genomes of Relative Wild Species

Cultivated species of plants were domesticated from their wild relatives. The use of molecular markers and the sequencing of the DNA of domesticated plants and closely related wild taxa have significantly increased our understanding of crop plant evolution. In addition to providing the most accurate measure of relatedness between domesticated taxa and putative wild ancestors, these molecular systematic studies contribute essential information to archaeologists relating to the geography of plant domestication, and they serve as independent tests of hypotheses for agricultural origins from archaeobotanical and other kinds of research.

The origin and differentiation of Asian cultivated rice (*O. sativa* L.) has been a controversial subject in terms of its ancestor and the phylogenetic relationships between subspecies. Sun et al. (2002) have analyzed and integrated the genetic diversity of the three different classes of genome to infer the phylogenetic relationships of cultivated rice and its wild ancestor. The recent studies revealed that *indica-japonica* differentiation occurred not only in the nuclear genome but also in the cytoplasmic genomes of common wild rice. This supports the hypothesis of diphyletic origin. However, the mitochondria of the *indica* type might be more primitive than that of the *japonica* type, indicating that the mitochondria of ancestor species domesticated into the *indica* type earlier than the *japonica* type. This evidence supports the hypothesis that the hsien rice *(indica)* was developed from the common wild rice in Southern China, and the keng rice *(japonica)* differentiated from Hsien later (Ting, 1957, 1961).

CONCLUDING REMARKS

Many different molecular techniques have been used to assess the genetic differentiation of nuclear, mitochondrial, and chloroplast genomes of plants. Information on their DNA variations has provided a comprehensive understanding of phylogenetic relationships among species and wider taxanomic groups. Some early conclusions were based upon analysis of only a few loci or regions of the genomes. It is important to integrate more recent information, including details of nuclear, mitochondrial, and chloroplast genomes, to reveal the systematic and phylogenetic relationships among species for plant conservation.

REFERENCES

Adams, K.L., Qiu, Y.L., Stoutemyer, M., and Palmer, J.D. (2002). Punctuated evolution of mitochondrial gene content: High and variable rates of mitochondrial gene loss and transfer to the nucleus during angiosperm evolution. *Proceedings of the National Academy of Sciences USA* 99, 9905-9912.

Arabidopsis Genome Initiative (2000). Analysis of the genome sequence of the flowering plant *Arabidopsis thaliana. Nature* 408, 796-815.

Baum, B.R. and Appels, R. (1992). Evolutionary change at the 5S DNA loci of species in the Triticeae. *Plant Systematics and Evolution* 183, 195-208.

Bennetzen, J.L. (2000). Comparative sequence analysis of plant nuclear genomes: Microcolinearity and its many exceptions. *Plant Cell* 12, 1021-1029.

Berg, O.G. and Kurland, C.G. (2000). Why mitochondrial genes are most often found in nuclei. *Molecular Biology and Evolution* 17, 951-961.

Bergthorsson, U., Adams, K.L., Thomason, B., and Palmer, J.D., (2003). Widespread horizontal transfer of mitochondrial genes in flowering plants. *Nature* 424, 197-201.

Breiman, A. (1987). Mitchondrial DNA diversity in the genera of *Triticum* and *Aegilops* revealed by Southern-blot hybridization. *Theoretical and Applied Genetics* 73, 563-570.

Chen, L., Liang, C.Y., Sun, C.Q., Jin, D.M., Jiang, T.B., Wang, B., and Wang, X.K. (2001). Comparison between AFLP and RFLP markers in detecting the diversity of rice *(Oryza sativa* L.). *Agricultural Sciences in China* 1, 713-719.

Chen, M., SanMiguel, P., de Oliveira, A.C., Woo, S.-S., Zhang, H., Wing, R.A., and Bennetzen, J.L. (1997). Microcolinearilty in sh2-homologous regions of maize, rice, and sorghum genomes. *Proceedings of the National Academy of Sciences USA* 94, 3431-3435.

Chen, W.B., Nakamura, I., Sato, Y.I., and Nakai, H. (1993). Distribution of deletion type in cpDNA of cultivated and wild rice. *Japanese Journal of Genetics* 68, 597-603.

Chen, W.B., Nakamura, I., Sato, Y.I., and Nakai, H. (1994). Indica-japonica differentiation in Chinese rice landraces. *Euphytica* 74, 195-201.

Cipriani, G., Fiori, A., Moroldo, M., De Pauli, P., Messina, R., and Testolin, R. (2003). Screening chloroplast, mitochondrial, and nuclear DNA sequences suitable for taxonomic studies in Actinidaceae. *Acta Horticulturae* 610, 337-342.

Dally, A.M. and Second, G. (1990). Chloroplast DNA diversity in wild and cultivated species of rice (genus *Oryza*, Section Oryza): Cladistic-mutation and genetic-distance analysis. *Theoretical and Applied Genetics* 80, 209-222.

Demesure, B., Comps, B., and Petit, R.J. (1996). Chloroplast DNA phylogeography of the common beech (*Fagus sylvatica* L.) in Europe. *Evolution* 50, 2515-2520.

Demesure, N. and Petit, R.J. (1995). A set of universal primers for amplification of polymorphic non-coding regions of mitochondrial and chloroplast DNA in plants. *Molecular Ecology* 4, 129-131.

Devos, K.M. and Gale, M.D. (2000). Genome relationships: The grass model in current research. *Plant Cell* 12, 637-646.

Doi, K., Yoshimura, A., Nakano, M., Iwata, N., and Vaughan, D.A. (1995). Phylogenetic study of a genome species of genus *Oryza* using nuclear RFLP. *Rice Genetics Newsletter* 12, 160-162.

Dunford, R.P., Kurata, N., Laruie, D.A., Money, T.A., Minobe, Y., and Moore, G., (1995). Conservation of fine-scale DNA marker order in the genomes of rice and the Triticeae. *Nucleic Acids Research* 23, 2724-2728.

Eckenrode,V.K. and Levings III, C.S. (1986). Maize mitochondrial genes. *In Vitro Cellular and Developmental Biology* 22, 169-176.

El Mousadik, A. and Petit, R.J. (1996). Chloroplast DNA phylogeography of the argan tree of Morocco. *Molecular Ecology* 5, 547-555.

Feng, Q., Zhang,Y., Hao, P., Wang, S., Fu, G., Huang, Y., Li, Y., Zhu, J., Liu, Y., and Hu, X. (2002). Sequence and analysis of rice chromosome 4. *Nature* 420, 316-320.

Flavell, R. (1980). The molecular characterization and organization of plant chromosomal DNA sequences. *Annual Review of Plant Physiology* 31, 569-596.

Foote, T., Roberts, M., Kurata, N., Sasaki, T., and Moore, G. (1997). Detailed comparative mapping of cereal chromosome regions corresponding to the Ph1 locus in wheat. *Genetics* 147, 801-807.

Gale, M.D. and Devos, K.M. (1998a). Comparative genetics after 10 years. *Science* 282, 656-659.

Gale, M.D. and Devos, K.M. (1998b). Comparative genetics in the grasses. *Proceedings of the National Academy of Sciences USA* 95, 1971-1974.

Goff, S.A., Rick, D., Lan, T.H., Presting, G., Wang, R., Dunn, M., Glazebrook, J., Sessions, A., Oeller, P., and Varma, H. (2002). Draft sequence of the rice genome (*Oryza sativa* L. ssp. *japonica*). *Science* 296, 92-100.

Gray, M.W. (1992). The endosymbiont hypothesis revisited. *International Review of Cytology* 141, 233-357.

Gray, M.W., Burger, G., and Lang, B.F. (1999). Mitochondrial evolution. *Science* 283, 1476-1481.

Hadis, J.T., Quinn, C.J., and Conn, B.J. (2003). Phylogeny of *Elatostema* (Urticaceae) using chloroplast DNA data. *Telopea* 10, 235-246.

Han, B. and Xue, Y. (2003). Genome-wide intraspecific DNA-sequence variations in rice: Current opinion. *Plant Biology* 6, 134-138.

Hiratsuka, J., Shimada, H., Whittier, R., Ishibashi, T., Sakamoto, M., Mori, M., Kondo, C., Honji, Y., Sun, C.R., Meng, B.Y., Li, Y.Q., Kanno, A., Nishizawa, Y., Hirai, A., Shinozaki, K., and Sugiura, M. (1989). The complete sequence of the rice (*Oryza sativa*) chloroplast genome: Intermolecular recombination between distinct tRNA genes accounts for a major plastid DNA inversion during the evolution of the cereals. *Molecular and General Genetics* 217, 185-194.

Huang, Y.H., Sun, C.Q., and Wang, X.K. (1996). Indica-japonica differentiation of chloroplast DNA in Chinese common wild rice populations. In *Monograph Origin and Differentiation of Chinese Cultivated Rice*. Beijing: China Agricultural University Press, pp. 166-170.

Ichikawa H., Hirai, A., and Katayama, T. (1986). Genetic analysis of Oryza species by molecular markers for chloroplast genomes. *Theoretical and Applied Genetics* 72, 353-358.

Ishii, T., Terachi, T., Mori, N., and Tsunewaki, K. (1993). Comparative study on the chloroplast, mitochondrial and nuclear genome differentiation in two cultivated rice species, *Oryza sativa* and *Oryza glaberrima,* by RFLP analysis. *Theoretical and Applied Genetics* 86, 88-96.

Ishii, T., Terachi, T., and Tsunewaki, K. (1986). Restriction endonuclease analysis of chloroplast DNA from cultivated rice species, *Oryza sativa* and *O. laberrima*. *Japanese Journal of Genetics* 61, 537-541.

Ishii, T., Terachi, T., and Tsunewaki, K. (1988). Restriction endonuclease analysis of chloroplast DNA from A-genome diploid species of rice. *Japanese Journal of Genetics* 63, 523-536.

Kanno, A., Watanabe, N., Nakamura, I., and Hirai, A. (1993). Variation in chloroplast DNA from rice (*Oryza sativa*): Differences between deletions mediated by short direct-repeat sequences within a single species. *Theoretical and Applied Genetics* 86, 579-584.

Kemble, R.J. and Shepard, J.F. (1984). Cytoplasmic DNA variation in a potato protoclonal population. *Theoretical and Applied Genetics* 69, 211-216.

Kikuchi, S., Satoh, K., Nagata, T., Kawagashira, N., Doi, K., Kishimoto, N., Ya-zaki, J., Ishikawa, M., Yamada, H., Ooka, H., et al. (2003). Collection, mapping, and annotation of over 28,000 cDNA clones from japonica rice. *Science* 301, 376-379.

Kilian, A., Chen, J., Han, F., Steffenson, B., and Kleinhofs, A. (1997). Towards map-based cloning of the barley stem rust resistance genes *Rpg1* and *rpg4* using rice as an intergenomic cloning vehicle. *Plant Molecular Biology* 35, 187-195.

Laurent, V., Risterucci, A.M., and Lanaud, C. (1993). Chloroplast and mitochondrial DNA diversity in *Theobroma cacao. Theoretical and Applied Genetics* 87, 81-88.

Leitch, I.J., Chase, M.W., and Bennett, M.D. (1998). Phylogenetic analysis of DNA C-values provides evidence for a small ancestor genome size in flowering plants. *Annals of Botany* 82, 85-94.

Luo, H., Coppenolle, B.V., Seguin, M., and Boutry, M. (1995). Mitochondrial DNA phylogenetic relationships in *Hevea brasiliensis. Molecular Breeding* 1, 51-63.

Mackenzie, S. and McIntosh, L. (1999). Higher plant mitochondria. *Plant Cell* 11, 571-586.

Maier, R.M., Neckermann, K., Igloi, G.L., and Kossel, H. (1995). Complete sequence of the maize chloroplast genome: Gene content, hotspots of divergence and fine tuning of genetic information by transcript editing. *Journal Molecular Biology* 251, 614-628.

Martin, W., Stoebe, B., Goremykin, V., Hansmann, S., Hasegawa, M., and Kowal-lik, K.V. (1998). Gene transfer to the nucleus and the evolution of chloroplasts. *Nature* 393, 162-165.

McIntyre, C.L., Winberg, B., Houchins, K., Appels, R., and Baum, B.R. (1992). Relationships between *Oryza species* (Poaceae) based on 5S DNA sequences. *Plant Systematics and Evolution* 183, 249-264.

Millen, R.S., Olmstead, R.G., Adams, K.L., Palmer, J.D., Lao, N.T., Heggie, L., Kavanagh, T.A., Hibberd, J.M., Gray, J.C., Morden, C.W., Calie, P.J., Jermiin, L.S., and Wolfe, K.H. (2001). Many parallel losses of *infA* from chloroplast DNA during angiosperm evolution with multiple independent transfers to the nucleus. *Plant Cell* 13, 645-658.

Moran, G.F., Smith, D., Bell, J.C., and Appels, R. (1992). The 5S RNA genes in *Pinus radiata* and the spacer region as a probe for relationships between *Pinus* species. *Plant Systematics and Evolution* 183, 209-221.

Neale, D.B. and Oerofe, R.R.S.R. (1989). Paternal inheritance of chloroplast and maternal inheritance of mitochondrial DNA in loblolly pine. *Theoretical and Applied Genetics* 77, 212-216.

Neale, D.B., Saghai-Maroof, M.A., Allard, R.W., Zhang, Q., and Jorgensen, R.A. (1986). Chloroplast DNA diversity in populations of wild and cultivated barley. *Genetics* 120, 1105-1110.

Nei, M. (1987). *Molecular Evolutionary Genetics.* New York: Columbia University Press.

Nei, M. and Li, W.H. (1979). Mathematical model for studying genetic variation in terms of restriction endonucleases. *Proceedings of the National Academy of Sciences USA* 76, 5269-5273.

Newton, K.J. (1988). Plant mitochondrial genomes: Organization, expression and variation. *Annual Review of Plant Physiology* 39, 503-532.

Notsu, Y., Masood, S., Nishikawa, T., Kubo, N., Akiduki, G., Nakazono, M., Hirai, A., and Kadowaki, K. (2002). The complete sequence of the rice (*Oryza sativa* L.) mitochondrial genome: Frequent DNA sequence acquisition and loss during the evolution of flowering plants. *Molecular Genetics and Genomics* 268, 434-445.

Palmer, J.D. and Herbon, L.A. (1988). Plant mitochondrial DNA evolves rapidly in structure, but slowly in sequence. *Journal of Molecular Evolution* 28, 87-97.

Parsons, B.J., Newbury, H.J., Jackson, M.T., and Ford-Lloyd, B.V. (1997). Contrasting genetic diversity relationships are revealed in rice (*Oryza sativa* L.) using different marker types. *Molecular Breeding* 3, 115-125.

Paterson, A.H., Bowersa, J.E., Burow, M.D., Draye, X., Elsikc, C.G., Jiang, C.X., Katsar, C.S., Lan, T.H., Lin, Y.R., Ming, R., and Wright, R.J. (2000). Comparative genomics of plant chromosomes. *Plant Cell* 12, 1523-1540.

Powell, W., Morgante, M., Andre, C., McNicol, J.W., Machray, G.C., Doyle, J.J., Tingey, S.V., and Rafalski, J.A. (1995). Hypervariable microsatellites provide a general source of polymorphic DNA markers for the chloroplast genome. *Current Biology* 5, 1023-1029.

Provan, J., Russell, J.R., Booth, A., and Powell, W. (1999). Polymorphic chloroplast simple sequence repeat primers for systematic and population studies in the genus *Hordeum*. *Molecular Ecology* 8, 505-511.

Provan, J., Soranzo, N., and Wilson, N.J., (1998). Gene pool variation in Caledonian and European Scots pine (*Pinus sylvestris* L.) revealed by chloroplast simple sequence repeats. *Proceedings of the Royal Society of London Series B* 265, 1697-1705.

Rice Chromosome 10 Sequencing Consortium (2003). In-depth view of structure, activity, and evolution of rice chromosome 10. *Science* 300, 1566-1569.

Sanjur, O.I., Piprrno, D.R., Andres, T.C., and Wessel-Beaver, L. (2002). Phylogenetic relationships among domesticated and wild species of *Cucurbita* (Cucurbitaceae) inferred from a mitochondrial gene: Implications for crop plant evolution and areas of origin. *Proceedings of the National Academy of Sciences USA* 99, 535-540.

Sano, Y. and Sano, R. (1990). Variation of the intergenic spacer region of ribosomal DNA in cultivated and wild rice species. *Genome* 33, 209-218.

Sasaki, T., Matsumoto, T., Yamamoto, K., Sakata, K., Baba, T., Katayose, Y., Wu, J., Niimura, Y., Cheng, Z., and Nagamura, Y. (2002). The genome sequence and structure of rice chromosome 1. *Nature* 420, 312-316.

Schmidt, R. (2002). Plant genome evolution lesions: From comparative genomics at the DNA level. *Plant Molecular Biology* 48, 21-37.

Song, K.M., Osborn, T.C., and Williams, P.H. (1988). *Brassica* taxonomy based on nuclear restriction fragment length polymorphisms (RFLPs): 1. Genome evolution of diploid and amphidiploid species. *Theoretical and Applied Genetics* 75, 784-794.

Song, K.M., Osborn, T.C., and Williams, P.H. (1990). *Brassica* taxonomy based on nuclear restriction fragment length poly morphisms (RFLPs) 3. Genome relationships in *Brassica* and related genera and the origin of *B. oleracea* and *B. rapa* (syn. *campestris*). *Theoretical and Applied Genetics* 76, 497-506.

Sun, C.Q., Wang, X.K., Li, Z.C., Yoshimura, A., and Iwata, N. (2001). Comparison on the genetic diversity of common wild rice (*Oryza rufipogon* Griff.) and cultivated rice (*O. sativa* L.) using RFLP markers. *Theoretical and Applied Genetics* 102, 157-162.

Sun, C.Q., Wang, X.K., Yoshimura, A., and Doi, K. (2002). Genetic differentiation for nuclear, mitochondrial and chloroplast genomes in common wild rice (*O. rufipogon* Griff.) and cultivated rice (*O. sativa* L.). *Theoretical and Applied Genetics* 104, 1335-1345.

Timothy, D.H., Levings, C.S., Pring, D.R., Cone, M.F., and Kermicle, J.L. (1979). Organelle DNA variation and systematic relationships in the genus *Zea:* Teosinte. *Proceedings of the National Academy of Sciences USA* 76, 4220-4224.

Ting, Y. (1957). The origin and evolution of cultivated rice in China. *Acta Agronomia Sinica* 8, 243-260 [in Chinese with English abstract].

Ting, Y. (1961). *Rice Crop Science in China*. Beijing: Agricultural Press.

Turmel, M., Otis, C., and Lemieux, C. (2002). The chloroplast and mitochondrial genome sequences of the charophyte *Chaetosphaeridium globosum:* Insights into the timing of events that restructured organelle DNAs with the green algal lineage that led to lands plants. *Proceedings of the National Academy of Sciences USA* 99, 11275-11280.

Udovicic, F., McFadden, G.L., and Ladiges, P.Y. (1995). Phylogeny of *Eucalyptus* and *Angophora* based on 5S rDNA spacer sequence data. *Molecular Phylogentics and Evolution* 4, 247-256.

Virk, P.S., Zhu, J., Newbury, H.J., Bryan, G.J., Jackson, M.T., and Ford-Lloyd, B.V. (2000). Effectiveness of different classes of molecular marker for classifying and revealing variation in rice (*Oryza sativa*) germplasm. *Euphytica* 112, 275-284.

Wang, Z.Y., Second, G., and Tanksley, S.D. (1992). Polymorphism and phylogenetic relationships among species in the genus Oryza as determined by analysis of nuclear RFLPs. *Theoretical and Applied Genetics* 83, 565-581.

Wang, Z.Y. and Tanksley, S.D. (1989). Restriction fragment length polymorphism in *Oryza sativa* L. *Genome* 32, 1113-1118.

Whitkus, R., Doebley, J., and Lee, M. (1992). Comparative genome mapping of sorghum and maize. *Genetics* 132, 1119-1130.

Wolf, K.H., Li, W.H., and Sharp, P.M. (1987). Rates of necleotide substitution vary greatly among plant mitochondrial, chloroplast, and nuclear DNAs. *Proceedings of the National Academy of Sciences USA* 84, 9054-9058.

Wu, J., Krutovskii, K.V., and Strauss, S.H. (1998). Abundant mitochondrial genome diversity, population differentiation and convergent evolution in pines. *Genetics* 150, 1605-1614.

Yu, J., Hu, S., Wang, J., Wong, G.K., Li, S., Liu, B., Deng, Y., Dai, L., Zhou, Y., Zhang, X., et al. (2002). A draft sequence of the rice genome (*Oryza sativa* L. ssp. *indica*). *Science* 296, 79-92.

Zhang, Q.F., Saghai, M.A., Lu, T.Y., and Shen, B.Z. (1992). Genetic diversity and differentiation of indica and japonica rice detected by RFLP analysis. *Theoretical and Applied Genetics* 83, 495-499.

Zhu, J., Gale, M.D., Quarrie, S., Jackson, M.T., and Bryan, G.J. (1998). AFLP markers for the study of rice biodiversity. *Theoretical and Applied Genetics* 96, 602-611.

Zhu, Z.F., Sun, Q., Jiang, T.B., Fu, Q., and Wang, X.K. (2001). The comparison of genetic divergences and its relationship to heterosis revealed by SSR and RFLP markers in rice (*Oryza sativa* L.). *Acta Genetica Sinica* 28, 738-746.

Chapter 11

Genomics and Plant Biodiversity Management

Robert J. Henry

INTRODUCTION

Genomics extends genetic studies by considering the contributions of all or many of the genes in a plant. The rapid advancement of DNA technologies has allowed analysis of the entire genomes of some plant species. Many traditional molecular genetics approaches are now considered also to fall within the scope of genomics. For example, mapping of plant genomes becomes genomics because much larger numbers or all of the genes in a plant are analyzed. Genomics provides powerful tools for plant conservation of both wild populations and domesticated plants. The analysis of plant genomes has revealed variations at many levels: the size of the genome (DNA content), number of genes, number of copies of each chromosome or gene (ploidy), abundance and nature of repetitive elements, the relative positions of genes in the genome (comparative mapping), and codons or specific bases. The analysis of plant DNA at any or all of these levels can provide information on which to base better management of genetic diversity.

VARIATION IN SIZES OF GENOMES

Plant genomes vary greatly in size, with a more than 1,000-fold range in the DNA content, among different flowering plant species (Wendel et al., 2002) and even among closely related species. The conservation of diversity in genome size may not be a key issue in plant

doi:10.1300/5546_11

genome conservation, and evidence suggests that expansion and contraction of genomes have occurred repeatedly during evolution.

NUMBER OF GENES

The number of genes in plant genomes is not well defined (Bennett, 1998). Estimates in rice are currently about 40,000 to 50,000. However, it is likely that gene number varies significantly among species. Genomics is likely to continue to provide more information on the diversity of genes and on the variations in the numbers of genes in plants. The conservation of any unique genes found in a species is an important target for conservation of plant gene diversity.

PLOIDY AND GENE DUPLICATION

The number of chromosomes in plants and the number of copies of each chromosome vary widely (Otto and Whitton, 2000). The evolutionary value of high levels of ploidy may be the redundancy that multiple copies of genes provide. Polyploidy may allow individual copies of genes to specialize. For example, different members of the gene family may be expressed in different tissues, allowing fine control of tissue specificity and developmental expression pattern. Polyploidy may be an important contributor to plant speciation (Levin, 2001). Methylation, which suppresses gene expression, may have an important role in the evolution of polyploid genomes (Hafiz et al., 2001).

REPETITIVE ELEMENTS

Highly repetitive elements are the major component of most genomes (Feschotte et al., 2002). These elements are transposable and of little conservation value, but may represent an important source of sequences in which new genes can evolve. Little is known about the abundance and distribution of these elements in most plant genomes (Meyers et al., 2001).

COMPARATIVE GENOMICS

The arrangements of genes within plant genomes can be highly conserved, especially between closely related species. Conserved genes are useful tools for analysis of genome evolution (Fulton et al., 2002). The conservation of gene order provides crucial evidence of evolutionary relationships and may be used as the basis of research. The functions of related (orthologous) genes can differ between species, providing a key basis of diversification of plant species (Shimamoto and Kyozuka, 2002).

BASE COMPOSITION AND CODON USAGE

Codon usage and the predominance of G and C rather than A and T are variable. A gradient in base composition from the start of genes and extending about 1,000 bp has been detected in the rice genome (Goff et al., 2002). Codon bias may be related to gene function, with housekeeping genes having a higher G+C content and genes with more specific expression patterns having a higher A+T content (Chiapello et al., 1998). Conservation of this diversity of base composition may be important for effective gene function.

UNDERSTANDING THE EVOLUTION OF PLANT GENOMES

The results of genomic studies may allow better management strategies that encourage continued evolution. A knowledge of relationships can guide plant improvement by identifying close relatives that may be able to interbreed. Discoveries of new genes may also be accelerated by an understanding of evolutionary relationships. Much of the divergence in gene sequences between plant species is probably of adaptive significance (Mitchell-Olds and Clauss, 2002).

OPPORTUNITIES

Genomics provides the large volumes of genetic data necessary for the development of cost-effectiveness techniques to promote biodiversity conservation. Genotyping of plants is a key tool in the analysis of options for critically endangered species. Conservation of diversity in more abundant species also requires an understanding of the genetic structure of the populations.

APPLICATIONS TO DATE

Genomics approaches have not been used extensively for plant conservation. Molecular methods can be helpful in defining phylogenetic relationships over large or small genetic distances, untangling population genetics, and better understanding the impact of domestication on genetic resources in some species. Identification of genotypes in collections is also an important use.

Genomics provides many new options for phylogenetic analysis. It enables researchers to choose the loci that have the appropriate levels of sequence polymorphism for the genetic distances being studied. The following examples are largely from systems studied by the author at the Centre for Plant Conservation Genetics at Southern Cross University.

A small number of genetic loci were used in genetic studies before genomic approaches were developed. Ribosomal genes were widely used in traditional plant phylogenetic analyses. The internal transcribed spacers (ITS) of the ribosomal genes (Dillon et al., 2001), the sequence of the 18S ribosomal gene (Graham et al., 1996), and the 5S ribosomal gene and the 5S spacer (Ko and Henry, 1996) have all been useful. Other genes that have been widely used include *adh* and *rbc*L. Single-copy genes simplify analysis by avoiding the complications associated with multiple-copy genes, such as the ribosomal genes. The granule-bound starch synthase (GBSSI) is a good example of a single-copy locus that has been useful in plant phylogenetic analysis (Mason-Gamer et al., 1998).

Chloroplast genes used in phyolgentic analysis include *mat*K, *ndh*F, and intergenic spacers (Goremykin et al., 1996). Stoke et al. (2001) used the J_{LA} region of the chloroplast to identify natural interspecific hybrids in *Eucalyptus*.

The entire sequence of the chloroplast genome is now available for a growing number of species. The whole genome nuclear sequences and the large expressed sequence tag (EST) databases allow the identification of much larger numbers of genetic loci.

The availability of a whole genome sequence has greatly simplified the analysis of rice genes using linked simple sequence repeat (SSR) or microsatellite markers (Cordeiro et al., 2002). Large-scale analyses have been conducted by applying SSR markers (Rossetto, Slade, et al., 1999) derived from enriched genomic libraries (Rossetto, McLaudan, et al., 1999). The EST data can also provide a readily available source of SSR markers (Holton et al., 2002). Analysis of grape ESTs identified SSR loci (Scott et al., 2000) that proved useful for wild grape relatives (Rossetto et al., 2002). Similar successes (Table 11.1) have been reported for other species such as sugarcane (Cordeiro et al., 2001) and the Myrtaceae (Rossetto et al., 2000). EST-derived SSRs are more easily transferable to other species, providing an important source of SSRs for analysis of genetic relationships in related plant species. EST data can also contain the sequences of economically or environmentally important genes (Sheldon et al., 2002). These genes may be the subjects of selection in nature or during domestication.

More recent research has focused on the smallest unit of genetic difference, the single nucleotide polymorphism (SNP), which is a single base change in the DNA sequence. SNP markers can be identified efficiently by analysis of EST databases (Kota et al., 2003). The SNPs apparent in sequence databases can be confirmed by resequencing by a range of techniques (Pacey-Miller and Henry, 2003). The flanking sequences surrounding SSRs may also provide a source of

TABLE 11.1. Transfer of SSR Between Species (Examples from Research at the Centre for Plant Conservation Genetics)

Plant family	Genera	Source of SSR	Reference
Myrtaceae	*Melaleuca, Backhousia, Eucalyptus, Callistemon*	*Melaleuca* (genomic)	Rossetto et al., 2000
Vitaceae	*Vitis, Cissus, Cayratia*	*Vitis* EST	Scott et al., 2000
Poaceae	*Saccharum, Erianthus Sorghum*	*Saccharum* EST	Cordeiro et al., 2001
	Triticum, Hordeum	Triticeae EST	Holton et al., 2002
Pinaceae	*Pinus* (soft and hard pines) (different subgenera)	*Pinus* (various)	Shepherd et al., 2002

SNPs (Arnold et al., 2002), a useful option for high throughput SNP discovery.

Analyses of the distribution of SNPs across genes can pinpoint regions of high abundance. Bundock and Henry (2004) determined the frequency of SNPs in the gene encoding the bifunctional amylase/subtilisin inhibitor from wild and domesticated barley (Figure 11.1). This and other studies have shown the greater abundance of SNP outside coding regions. These studies also identify recombinations within genes and demonstrate the variations in recombination frequencies in different parts of the genome.

CHALLENGES FOR THE FUTURE

The main challenge in applying genomics to plant conservation is the large number of plant species that exist and the relatively small number for which much genomic data are available. The less expensive and higher throughput DNA methods will reduce the costs associated with application of these techniques to new species. The new technologies for massively parallel analysis of plant genes such as microarray analysis (Aharoni and Vorst, 2001) will accelerate these developments.

Reticulate evolution can confuse phylogenetic analysis based on some loci and probably restricts the analysis of plant relationships using larger numbers of genetic loci. Movement of genes between the

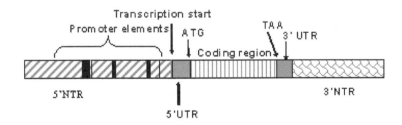

FIGURE 11.1. Analysis of the Polymorphism in a Gene. *Note:* The *asi* gene from barley was analyzed by Bundock and Henry (2003). Polymorphism was found to be much higher in the upstream and downstream sequences than in the coding region. One single nucleotide polymorphism (SNP) per 75 bp was found in the transcribed region, with ten SNP in the coding region, none in the 5' UTR (untranslated region), and one in the 3' UTR. The greatest SNP frequency (one per 16 bp) was found in the 3' NTR (nontranscribed region). *Source:* Adapted from Bundock and Henry, 2004.

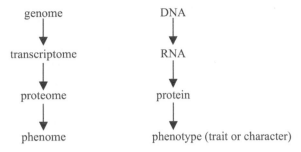

FIGURE 11.2. Levels of Analysis of Plant Diversity.

chloroplast, mitochondrial, and nuclear genomes can also complicate the analysis of some loci.

The cost of collection and documentation of DNA from plant populations is a limiting factor in the application of genomics analysis. These costs may greatly exceed those associated with the DNA analyses in many studies. The availability of large collections of plant DNA in repositories such as DNA banks may help overcome this limitation in the future.

Plant diversity can be assessed at many levels other than the genomic or DNA level (Figure 11.2). Analysis of RNA, protein, and phenotype are also used to measure biodiversity. Phenotypic data are the basis of traditional morphological taxonomy and diversity analysis. Analysis at the RNA and protein levels are not yet as well developed as DNA analysis. They also have the disadvantage of being influenced by the environment and by the plant genotype. However, the large-scale analysis of expressed genes has resulted in rapid expansion of EST data, which are a valuable source of sequence information for plant research. The rapidly advancing field of proteomics is also anticipated to contribute to plant diversity analysis.

REFERENCES

Aharoni, A. and Vorst, O. (2001). DNA microarrays for functional plant genomics. *Plant Molecular Biology* 48, 99-118.

Arnold, C., Rossotto, M., McNally, J., and Henry, R.J. (2002). The potential application of SSRs characterised for grape *(Vitis vinifera)* to conservation studies in Vitaceae. *American Journal of Botany* 89, 22-28.

Bennett, M.D. (1998). Plant genome values: How much do we know? *Proceedings of the National Academy of Sciences of the United States of America* 95, 2011-2016.

Bundock, P. and Henry, R.J. (2004). Single nucleotide polymorphism, haplotype diversity and recombination in the *asi* gene of barley. *Theoretical and Applied Genetics* 109, 543-551.

Chiapello, H., Lisacek, F., Caboche, M., and Henaut, A. (1998). Codon usage and gene function are related in sequences of *Arabidopsis thaliana*. *Gene* 209, 1-38.

Cordeiro, G., Casu, R., McIntyre, C.L., Manners, J.M., and Henry, R.J. (2001). Microsatellite markers from sugarcane (*Saccharum* spp.) ESTs. *Plant Science* 160, 1115-1123.

Cordeiro, G.M., Christopher, M.J., Henry, R.J., and Reinke, R.F. (2002). Identification of microsatellite markers for fragrance in rice by analysis of the rice genome. *Molecular Breeding* 9, 245-250.

Dillon, S.L., Lawrence, P.K., and Henry, R.J. (2001). The use of ribosomal ITS to determine phylogenetic relationships within *Sorghum*. *Plant Systematics and Evolution* 230, 97-110.

Feschotte, C., Jiang, N., and Wesseler, S.R. (2002). Plant transposable elements: Where genetics meets genomics. *Nature Reviews Genetics* 3, 329-341.

Fulton, T.M., Van der Hoeven, R., Eannetta, N.T., and Tanksley, S.D. (2002). Identification, analysis, and utilization of conserved ortholog set markers for comparative genomics an higher plants. *Plant Cell* 14, 1457-1467.

Goff, S.A., Ricke, D., Lan, T.H., Presting, G., Wang, R., Dunn, M., Glazebrook, J., Sessions, A., Oeller, P., Varma, H., et al. (2002). A draft sequence of the rice genome (*Oryza sativa* L. ssp japonica). *Science* 296, 92-100.

Goremykin, V., Bobrova, V., Pahnke, J., Troitsky, A., Antanov, A., and Martin, W. (1996). Noncoding sequences from the slowly evolving chloroplast inverted repeat in addition to *rbc*L data do not support Gnetalean affinities of angiosperms. *Molecular Biology and Evolution* 13, 383-396.

Graham, G.C., Henry, R.J., Godwin, I.D., and Nikles, G.D. (1996). Phylogenetic position of hoop pine *(Araucaria cunninghamii). Australian Systematic Botany* 9, 893-902.

Hafiz, I.A., Anjum, M.A., Grewal, A.G., and Chaudhary, G.A. (2001). DNA methylation—an essential mechanism in plant molecular biology. *Acta Physiologiae Plantarum* 23, 491-499.

Holton, T.A., Christopher, J., McLure, L., Harker, N., and Henry, R.J. (2002). Identification of polymorphic SSR markers from expressed genes sequences of barley and wheat. *Molecular Breeding* 9, 63-71.

Ko, H.L. and Henry, R.J. (1996). Specific 5S ribosomal RNA primers for plant species identification in admixtures. *Plant Molecular Biology Reporter* 14, 33-43.

Kota, R., Rudd, S., Facius, A., Kolesov, G., Theil, T., Zhang, H., Stein, N., Mayer, K., and Graner, A. (2003). Snipping polymorphisms from the EST collections in barley (*Hordeum vulgare* L.). *Molecular Genetics and Genomics* 270, 24-33.

Levin, D.A. (2001). 50 years of plant speciation. *Taxon* 50, 69-91.

Mason-Gamer, R.J., Weil, C.F., and Kellog, E.A. (1998). Granule-bound starch synthase: Structure, function, and phylogenetic utility. *Molecular Biology and Evolution* 15, 1658-1673.

Meyers, B.C., Tingley, S.V., and Morgante, M. (2001). Abundance, distribution, and transcriptional activity of repetitive elements in the maize genome. *Genome Research* 11, 1660-1676.

Mitchell-Olds, T. and Clauss, M.J. (2002). Plant evolutionary genomics. *Current Opinion in Plant Biology* 5, 74-79.

Otto, S.P. and Whitton, J. (2000). Polyploid incidence and evolution. *Annual Review of Genetics* 34, 401-437.

Pacey-Miller, T. and Henry, R.J. (2003). SNP detection in plants using a single stranded pyrosequencing protocol with a universal biotinylated primer. *Analytical Biochemistry* 317, 165-170.

Rossetto, M., Harris, F.L.C., McLauclan, A., Henry, R.J., Baverstock, P., and Lee, L.S. (2000). Interspecific amplification of tea tree (*Melaleuca alternifolia* Cheel) (Myrtaceae). *Australian Journal of Botany* 48, 367-373.

Rossetto, M., McLauclan, A., Harriss, F., Henry, R.J., Lee, L.S., Baverstock, P., Maguire, T., and Edwards, K. (1999). Abundance and polymorphism of microsatellite markers in tea tree (*Melaluca alternifolia* Myrtaceae). *Theoretical and Applied Genetics* 98, 1091-1098.

Rossetto, M., McNally, J., and Henry, R.J. (2002). Evaluating the potential of SSR flanking regions for examining taxonomic relationships in Vitaceae. *Theoretical and Applied Genetics* 104, 61-66.

Rossetto, M., Slade, R.W., Baverstock, P., Henry, R.J., and Lee, L.S. (1999). Microsatellite variation and analysis of genetic structure in tea tree (*Melaluca alternifolia* Myrtaceae) *Molecular Ecology* 8, 633-643.

Scott, K.D., Eggler, P., Seaton, G., Rossotto, M., Ablett, E.M., Lee, L.S., and Henry, R.J. (2000). Analysis of SSRs derived from grape ESTs. *Theoretical and Applied Genetics* 100, 723-726.

Sheldon, D., Leach, D., Baverstock, P., and Henry, R.J. (2002). Isolation of genes involved in secondary metabolism from *Melaleuca alternifolia* (Cheel) using expressed sequence tags (ESTs). *Plant Science* 162, 9-15.

Shepherd, M., Cross, M., Maguire, T., Dieters, M., Williams, C.G., and Henry, R.J. (2002). Transpacific microsatellites for hard pines. *Theoretical and Applied Genetics* 104, 819-827.

Shimamoto, K. and Kyozuka, J. (2002). Rice as a model for comparative genomics of plants. *Annual Review of Plant Biology* 53, 399-419.

Stoke, R.L., Shepherd, M., Lee, D.J., Nikles, G., and Henry, R.J. (2001). Natural inter-subgeneric hybridisation between *Eucalyptus acmenoides* Schauer and *Eucalyptus coleziana* F. Muell (Myrtaceae) in southeast Queensland. *Annals of Botany* 88, 563-570.

Wendel, J.F., Cronn, R.C., Johnston, J.S., and Price, H.J. (2002). Feast and famine in plant genomes. *Genetica* 115, 37-47.

Index

AFLP, 134
Amaranth, 70
Ananas comosus, 13
Apple, 17
Arabidopsis, 152
Arachis, 20
Armoracia, 20
ATCFC, 39, 44, 45
AusPGRIS (National Australian Plant
    Genetic Resource Information
    System), 52
Australian Tropical Crops and Forages
    Collection (ATCFC), 39,
    44, 45
Avocado, 38

Banana, 17
Benefit sharing, 88
6-benzyladenine, 17
Biodiversity management, 167
Botanic garden, 11, 5, 75
Brassica, 150
Breadfruit, 38

Cacao, 13
Cajanus, 41, 44
Cajanus cajan, 41
Cajanus lanuginosis, 45
Carrot, 20
Cassava, 17
CATIE (Centro Agronomico
    Tropical de Investigacion
    y Ensenanza), 22

CBD (Convention on Biological
    Diversity), 7, 26, 78, 89, 105,
    114
Cedrela fissilis, 16
Centro Agronomico Tropical de
    Investigacion y Ensenanza
    (CATIE), 22
Centro Internactional de Agricultura
    Tropical (CIAT), 17
Centro Internacional de la Papa
    (CIP), 21
Chaetosphaeridium globosum, 153
Chinese yam, 20
Chloroplast, 149
CIAT (Centro Internactional de
    Agricultura Tropical), 17
CIP (Centro Internacional de la
    Papa), 21
CITES (Convention on International
    Trade in Endangered
    Species), 78, 88
Citrus, 20, 23
Citrus, 22, 24
Citrus sinensis, 20
Cocoa, 38
Coconut, 13
Cocos nucifera, 13, 23
Coffea
    arabica, 13, 23
    congensis, 17
Coffee, 13, 38
Collecting itinerary, 43
Collecting mission, 39
Collecting seed, 46
Collecting site, 50
Collection priorities, 40
Colocasia esculenta, 13

PLANT CONSERVATION GENETICS

_____in hardbound at $49.95 (ISBN-13: 978-1-56022-996-4; ISBN-10: 1-56022-996-9)

_____in softbound at $34.95 (ISBN-13: 978-1-56022-997-1; ISBN-10: 1-56022-997-7)

Or order online and use special offer code HEC25 in the shopping cart.

COST OF BOOKS_____

☐ **BILL ME LATER:** (Bill-me option is good on US/Canada/Mexico orders only; not good to jobbers, wholesalers, or subscription agencies.)

☐ Check here if billing address is different from shipping address and attach purchase order and billing address information.

POSTAGE & HANDLING_____
(US: $4.00 for first book & $1.50 for each additional book)
(Outside US: $5.00 for first book & $2.00 for each additional book)

Signature_____

SUBTOTAL_____

☐ **PAYMENT ENCLOSED: $_____**

IN CANADA: ADD 7% GST_____

☐ **PLEASE CHARGE TO MY CREDIT CARD.**

STATE TAX_____
(NJ, NY, OH, MN, CA, IL, IN, PA, & SD residents, add appropriate local sales tax)

☐ Visa ☐ MasterCard ☐ AmEx ☐ Discover
☐ Diner's Club ☐ Eurocard ☐ JCB

Account # _____

FINAL TOTAL_____
(If paying in Canadian funds, convert using the current exchange rate, UNESCO coupons welcome)

Exp. Date_____

Signature_____

Prices in US dollars and subject to change without notice.

NAME_____

INSTITUTION_____

ADDRESS_____

CITY_____

STATE/ZIP_____

COUNTRY_____ COUNTY (NY residents only)_____

TEL_____ FAX_____

E-MAIL_____

May we use your e-mail address for confirmations and other types of information? ☐ Yes ☐ No
We appreciate receiving your e-mail address and fax number. Haworth would like to e-mail or fax special discount offers to you, as a preferred customer. **We will never share, rent, or exchange your e-mail address or fax number.** We regard such actions as an invasion of your privacy.

Order From Your Local Bookstore or Directly From
The Haworth Press, Inc.
10 Alice Street, Binghamton, New York 13904-1580 • USA
TELEPHONE: 1-800-HAWORTH (1-800-429-6784) / Outside US/Canada: (607) 722-5857
FAX: 1-800-895-0582 / Outside US/Canada: (607) 771-0012
E-mail to: orders@haworthpress.com

For orders outside US and Canada, you may wish to order through your local sales representative, distributor, or bookseller.
For information, see http://haworthpress.com/distributors

(Discounts are available for individual orders in US and Canada only, not booksellers/distributors.)

PLEASE PHOTOCOPY THIS FORM FOR YOUR PERSONAL USE.
http://www.HaworthPress.com   BOF06